Computer Game Development

Edited by Branislav Sobota

Published in London, United Kingdom

Computer Game Development
http://dx.doi.org/10.5772/intechopen.97983
Edited by Branislav Sobota

Contributors
Diana R. Sanchez, Amanda Rueda, Leila Jimeno Jimènez, Mahsa Norouzi Nargesi, Sarah Fachada, Daniele Bonatto, Mehrdad Teratani, Gauthier Lafruit, Bruce Campbell, Dario Liberona, Aravin Kumaresan, Lionel Valenzuela, Giovanny Tarazona, Branislav Sobota, Marián Hudák, Emília Pietriková, Miriama Mattová, Ingrid Nery Mendes, Otavio Noura Teixeira, Maicon de Araújo Nogueira, Filipe Valente Mendes, Viviane Almeida dos Santos, John Allison, Anita Alaere Bala

First published in London, United Kingdom, 2022 by IntechOpen
IntechOpen is the global imprint of INTECHOPEN LIMITED, registered in England and Wales, registration number: 11086078, 5 Princes Gate Court, London, SW7 2QJ, United Kingdom

British Library Cataloguing-in-Publication Data
A catalogue record for this book is available from the British Library

Additional hard and PDF copies can be obtained from orders@intechopen.com

Computer Game Development
Edited by Branislav Sobota
p. cm.
Print ISBN 978-1-80355-564-5
Online ISBN 978-1-80355-565-2
eBook (PDF) ISBN 978-1-80355-566-9

Meet the editor

Branislav Sobota obtained an MSc with honours from the Department of Computers and Informatics, Faculty of Electrical Engineering and Informatics, Technical University of Košice, Slovakia in 1990. He defended his Ph.D. in 1999 and habilitation thesis in the field of virtual reality and computer graphics in 2008. He is an associate professor at the Department of Computers and Informatics, Technical University of Košice. His scientific research focuses on computer graphics, parallel computing, virtual reality, and related technologies.

Contents

Preface

The computer games industry is continuously growing and attracting new players and game developers. At the end of 2021, the value of the gaming industry was about 200 billion US dollars. All industries, including the computer gaming industry, are under constant pressure to keep up with the latest technological developments. Thus, computer game developers have no choice but to adopt these new technologies, otherwise, they risk being left behind.

The growing performance of hardware has transformed the visions of game designers and developers into reality and led to better game mechanics, graphics, storytelling, and processing. Computer games provide players with hundreds of hours of fun with well-crafted and fully interactive storytelling with lots of different endings. What's even more fun is sharing the gaming experience with friends. Multiplayer games bring together players from all over the world, allowing them to work with friends to fulfill game goals or to show their skills in a game against other players.

Computer games and the gamification of some human activities on various platforms have become an important part of our daily lives. It is very likely that today's young person has played at least one computer game during their lifetime; more people are playing games than ever before. Computer games are usually the first communication step between a human and a computer or game console or smartphone.

This book describes some modern approaches, procedures, algorithms, and devices in the field of computer games. It discusses classifications and types of games, technical means of game systems, game consoles, and the newest tools based on extended reality (XR) technologies. It also discusses visualization and 3D computer graphics, game cores (engines) and platforms, genre selection, proper object, procedural, graphic as well as story design, and trends in game design. The book presents formalizations, algorithms, and mathematical tools for game theory. There is a focus on the application of virtual reality and related technologies to modern game user interfaces. Lastly, the book examines the sociological impacts of computer games and their development.

I would like to express my sincere gratitude to all the authors and co-authors for their contributions. The successful completion of this book has been the result of the cooperation among many people. I would especially like to thank Publishing Process Manager, Ms. Martina Ivancic, at IntechOpen for her support during the publishing process.

Branislav Sobota
Technical University of Košice,
Košice, Slovakia

Chapter 1

Collaborative XR Systems and Computer Games Development

Branislav Sobota, Marián Hudák and Emília Pietriková

Abstract

Computer games and, of course, the development associated with them have been in the spotlight for many years; recently, also with technologies, such as virtual reality or extended reality. The main part of this chapter presents the classification of collaborative XR systems, the concept of the major application architectures and the consistency models of shared virtual game environments. The next subchapter briefly deals with the sharing of property ownership. The mentioned concepts and examples used in the chapter are implemented in many works and projects using a collaborative environment (also in gamified form) developed in the laboratory LIRKIS, the home laboratory of the authors. The knowledge presented in this chapter may provide tips and inspiration for some other game projects, and practical and useful notes on the advantages or disadvantages of some systems will be interesting and useful.

Keywords: extended reality, virtual reality, mixed reality, augmented reality, web-based collaborative environments

1. Introduction

The development of computer games, the gaming industry in general, is "the driving force" in many ways behind the development of many technologies. From this point of view, perhaps the most important part of a computer game is its core. Its performance also determines the performance of the game itself. However, from the point of view of playability and game immersion, the interface of the computer game is also extremely important. Here is a very modern usage of just new technologies, one of which is virtual reality (VR) and its related technologies. Virtual reality interfaces represent a wide range of devices, applications, and ways to connect the user to the target game system. Recently, the current development of VR interfaces and systems has progressed in improving human–computer interaction (HCI).

In fact, there are three basic technological bases of systems that are used as a standard in accessing and interacting with users with three-dimensional environments or computer game objects. Each of these systems can work with a virtual context, but it differs in the way of displaying and controlling.

- Virtual reality (VR) is the first system with a fully immersive virtual environment. The basic definition says: "Virtual-reality system represents interactive computer system created an illusion of synthetic (virtual) space in real time based on total simulation within the environment of close relation human – computing system." VR systems provide a better experience and they are more interactive, but the complexity of their implementation is greater. VR subsystems are mainly divided according to the senses that affect: Visualization subsystem, acoustic subsystem, kinetic and statokinetic subsystem, touch and contact subsystem, and other senses (e.g., a smell, a taste, and a sensitivity to pheromones). There is not much reason to implement some senses (which are commonly perceived in the real world) in the virtual world. For example, the taste is one such sensation at present. Through VR systems, the immersion of the user in computer games is controlled mainly at three standard levels: Visual, audio, and touch. Each of these levels provides a relevant way of feedback depending on the context of the virtual (gaming) environment.

- Mixed reality (MR) represents another system focused mainly on the virtual objects synthetization into the user's physical space. Immersion levels in MR systems are identical to VR systems. However, the main difference in the way of interaction is the control of the synthesized (virtual) objects by means of gestural inputs based on the user's hand sensing. Due to the detection of physical space, MR systems are extended by spatial recognition technology. Through this technology, it is possible to achieve detailed observation of synthesized (virtual) objects. MR-based computer games represent an extremely innovative way. Virtual game objects are embedded in the real world and the experience of the game by players can be very emotional.

- Augmented reality (AR) is a system closely related to the recognition of information in an image due to the placement of virtual (three-dimensional) objects. Although AR systems use the same level of immersion as VR and MR systems, a difference is in the processing of visual feedback. This is primarily tied to the image information, which must be consistent and recognizable to achieve the desired interaction. Collaborative virtual reality systems are created to share user interaction and common virtual space. The main task of these systems is to provide a shared environment to multiple users in real time, as well as to ensure the sharing of interaction and communication. Despite the present technological advances and a significant improvement in the immersive properties of VR, MR, and AR systems, there is a need to unify their software compatibility. The solution of this problem is to make a common application platform available for different types of systems, regardless of their differences and methods of interaction.

As mentioned in Ref. [1], the existing VR, MR, and AR systems are beginning to merge under the area of innovative extended reality (XR), where "X" refers to the variable use of any of them (**Figure 1**). Thus, the current development of independent application platforms approaches the unification of these systems to the level of XR. The unification of VR, MR, and AR systems into the XR group creates a high presumption of improving user interaction and the level of immersion in a computer game.

According to Beglund [2], XR usage focuses mainly on interaction via wearable devices or sensory inputs. Huzaifa [3] highlights the potential of the XR interface in using hand gesturing inputs, which he considers necessary to achieve a fully natural interaction. XR extensions to support the web platform are also entering the realm of virtual collaboration. Today, the WebXR standard makes it possible to access XR devices via web browsers. This reinforces the creation of web systems with a wide

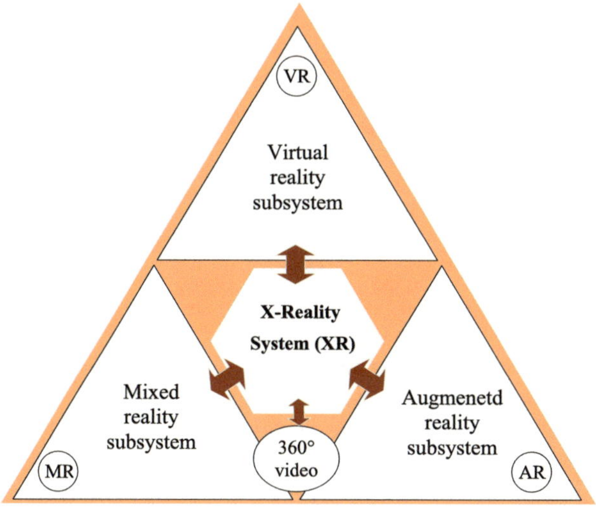

Figure 1.
XR system.

range of interactions and inputs [4]. Seo [5] considers WebXR to be widely applicable mainly to support a wide range of interaction techniques. In this way, it is possible to extend the web application level for different purposes. According to Anthes [6], the use of web technologies for the VR application sphere is necessary, especially due to the current hardware and software diversity. This is also confirmed by Wang [7], who justifies the need to implement a web platform for the use of heterogeneous VR devices. The emergence of web VR applications has contributed to the expansion and globalization of collaborative virtual environments, including gaming environments. Paiva [8] attributes high potential to the deployment of web collaborative environments for the purposes of education and training, where it positively evaluates the benefits of sharing user/player interaction.

Using web platforms to share virtual environments and user interaction is possible, thanks to a high level of network infrastructure and technologies. In this way, systems are rapidly formed to support global virtual collaboration, where players can access a common virtual gaming space. The main advantage of the availability of such a system is the ability to connect players regardless of their actual geographical location in the world. Another advantage results from the refactoring of functions and modules of the web system. This refactoring is significantly faster and more efficient than in standard VR, MR, and AR applications. Despite these advantages, there are limitations to these systems that need to be solved.

The globalization of web systems increases the assumption of their use by a large group of users. This increases the probability that users will access the web system by using different types of devices and operating systems. In this case, there is a growing need to improve the adaptive features of the systems and their user interfaces so that they can provide the same control functions for different types of hardware and software platforms simultaneously according to the capabilities of individual players.

In collaborative virtual reality systems, multiuser interaction is shared in real time. This is necessary for many games. Given the context of a shared environment, there are ways to manage the interaction by which groups of players together access shared objects. As these systems continue to expand with different types of interactions and functions, there is a growing need to improve the management of interaction techniques. Typical examples of extensions are physical and simulation subsystems or hardware components and control elements. The main problem occurs when multiple users/players use concurrent-subsystem-generated properties of shared objects or environments. In this case, a collision in the calculations may occur, and the shared interaction or virtual objects will be inconsistent. In the case of sharing the physical properties of objects, their chaotic behavior is often observable, while their consistency is different for each user at the same time.

Synchronization of shared interactions is highly efficient when a client–server network architecture is used. The sharing of interaction between players (clients) is controlled by a centralized computing node (server). The server provides information about the state of the virtual gaming environment for all clients in real time. The centralized compute node obtains data about each user's interactions. Then, it sends them to all members of the collaborative game group. The problem of sharing interactions is often associated with player-side resources/devices performance. These fully affect the rendering of the virtual environment, the processing of data from the server, and the control of the player's interaction. Insufficient computing performance of the user resources/devices is manifested by latency, which reduces the quality of collaboration. Therefore, it is necessary to focus on solving this problem area.

In the following subchapters of this chapter, the collaborative XR systems classifications are presented, followed by the concepts of the main application architectures and virtual game environment sharing consistency models. In conclusion, we briefly touch on object ownership sharing. The mentioned concepts and examples used in the chapter have been implemented in many works and projects (some examples are presented in the chapter too) using a collaborative environment (in a gamified form too) in the LIRKIS laboratory. The LIRKIS laboratory is a laboratory at the authors' home institution (Department of Computers and Informatics, Faculty of Electrical Engineering and Informatics, Technical University of Košice). The authors hope that readers will find here many tips and inspirations for their own work and they will also use practical notes on the advantages or disadvantages of individual systems.

2. Classification of collaborative XR systems according to availability

User collaboration is significantly affected by the availability of systems that provide a shared virtual environment. In fact, the availability of collaborative virtual environment systems is classified according to the local or global access of users/players, as well as their physical or remote (virtual) presence.

Collaborative XR systems are divided depending on availability into the following categories:

- Locally shared systems with centralized access,
- Locally shared systems with separate access,
- Globally shared systems.

Each of these categories has specific properties related to user access to the target system, the sharing of users' interactions, and the resources needed for a possible gaming environment implementation.

2.1 Locally shared systems with centralized access

Locally shared systems are related to the physical presence of players within a shared game space. Sharing the local environment by centralized players' access is equivalent to using a virtual cave system [9] (**Figure 2**). It contains a physically accessible environment with the possibility of players' group full immersion in real time. According to Nguyen [10], in a virtual cave system, the way of natural interaction between users is highly efficient and realistic, without the need to implement virtual avatars. Users cooperate in a jointly physically shared space. They are here realistically visible to each other [11]. According to Back [12], communication between users is in a natural way, without the need for external technology to share communication. In this case, any latency in users' communication with each other is excluded.

When using a locally available system with centralized user access, there are two basic types of interfaces (**Figure 2**), which are as follows:

- *User interfaces providing in-game player's interaction* with a shared virtual game environment [13]. They represent input subsystems for the control of the virtual

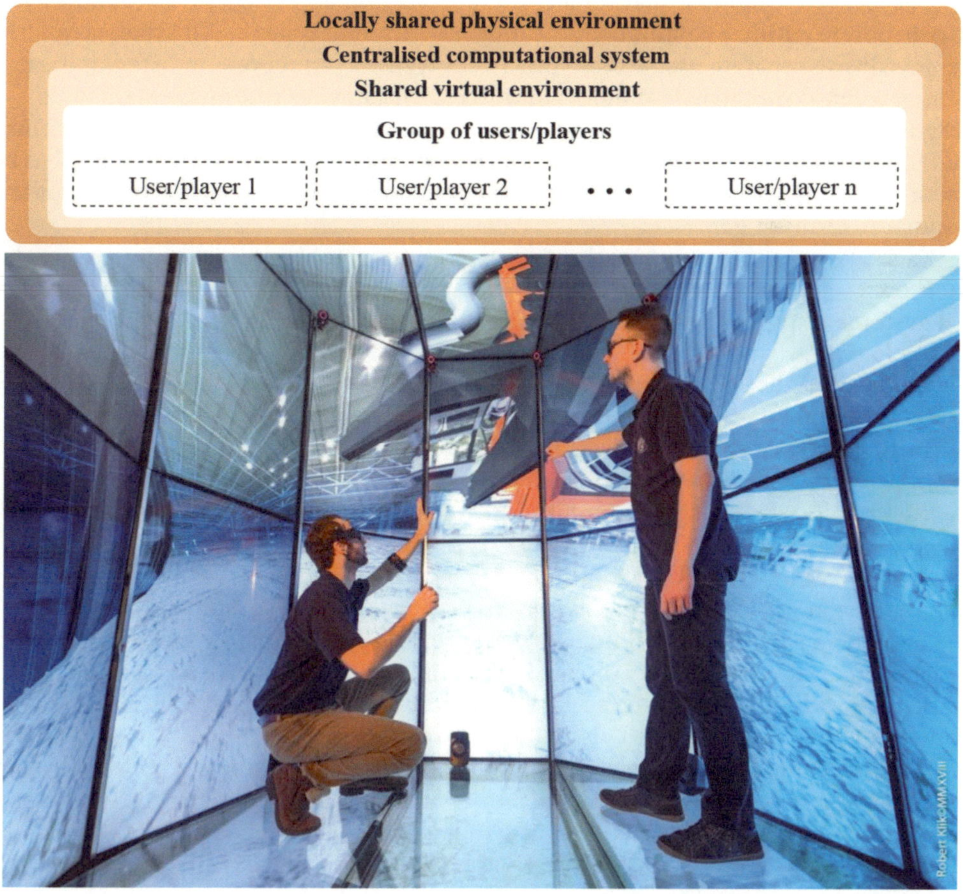

Figure 2.
Locally shared environment with centralized user access (LIRKIS CAVE virtual cave environment).

environment (e.g. touch interfaces, haptic controllers, sensor devices or gesticulation inputs).

- *Application interfaces*, the purpose of which is to obtain and process players' interactions and then submit them as feedback (visual, audio, and haptic). These interfaces provide program control between players' inputs and their subsequent processing into system outputs [14].

On the other hand, there are several limitations closely related, in particular, to the physical dimensions of such a system as well as to the multiuser interaction. In the case of virtual cave systems, the number of users collaborating/playing in real time is strictly limited. Another problem is the significant limitation of scene rendering for each user individually, mainly due to the use of stereoscopic visualization. The increasing number of users significantly multiplies the demands on the computing system performance due to the need to monitor their activity and movement simultaneously.

2.2 Locally shared systems with separate access

The local environment sharing using a separate user approach is the second way to collaborate within a common physical space. Different types of XR systems can be deployed with separate user access [15]. These systems can represent different data helmet technologies that communicate with each other via a communication node (**Figure 3**). The communication node creates a connection between separate systems, and it ensures the distribution of data. By default, the communication is controlled by using a local area network or another wireless standard. Nowadays, this type is quite popular in terms of games.

Each player is able to access the shared gaming virtual environment using their own separate system [16]. The most suitable for games is the deployment of XR systems using wireless technology, as it frees players from bonding to cable technology.

Unlike virtual cave systems, visual interaction is needed in this case [17], as users/players are not naturally visible to each other (unless MR systems are used). According to Horst et al. [18], users need to be replaced with virtual avatars. Then users/players in the virtual environment can communicate visually via these avatars. Furthermore, it is necessary to monitor the movement of users/players to avoid their physical collision during collaboration [19]. Today's XR technologies mostly include integrated free movement tracking subsystems that do not require the introduction of external technologies. The physical movement of the player is sensory processed by the integrated subsystem, and then, it is sent to the shared game virtual environment.

User's/player's interaction is created depending on the hardware technology used. Haptic and gestural input interfaces are most often used. The players are able to control and manipulate shared virtual objects with them.

Separate computing systems usage as, for example, data helmets give users/players much more freedom of movement and work with the broad context of the virtual game environment [20]. In this case, users/players move independently of each other, and they perform different types of activities simultaneously [21]. The group of collaborating users/players can be much larger than in virtual cave systems due to the use of separate computing systems.

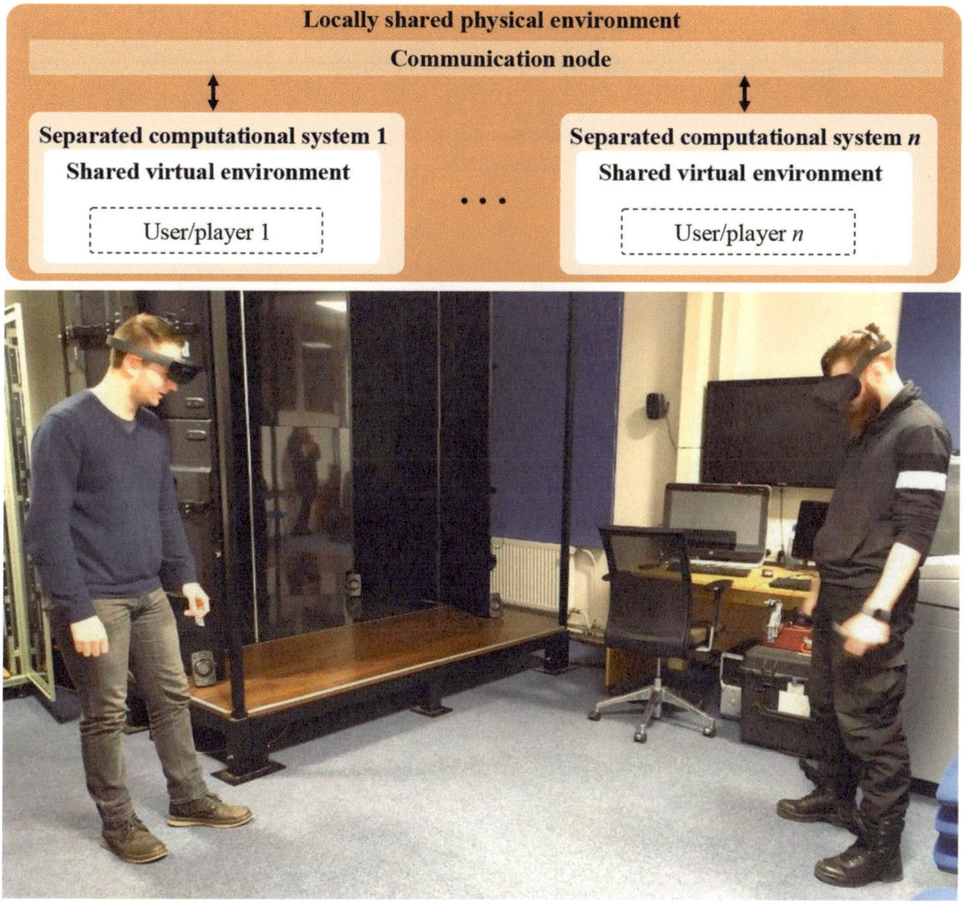

Figure 3.
Locally shared environment with separate user access (LIRKIS DCI FEEI TU Košice).

2.3 Globally shared systems

Sharing virtual environments with global access is mainly based on the resources of the internet and web technologies. According to Du et al. [22], by sharing globally available collaborative environments, it is possible to reach a large group of users for remote collaboration. This, of course, is also expanding the gaming business.

Global collaborative virtual environments use the computer network infrastructure (**Figure 4**). The users/players can interact with each other via this infrastructure. Instead of local physical space, globally shared space is used. Each user/player is able to access this space via a separate computing system [23]. Users/players connect to the server node via the network infrastructure, within which they communicate and interact with other users/players or game environments in real time [24]. Independence from the physical space of the user/player is the main advantage of these environments.

Globally shared environments are highly efficient in terms of sharing large-scale users/players activity as well as multipurpose deployments [9]. The most common problem in global virtual environment sharing is platform independence [25]. It is as a result of globalization that the demand for multiplatform compatibility is growing,

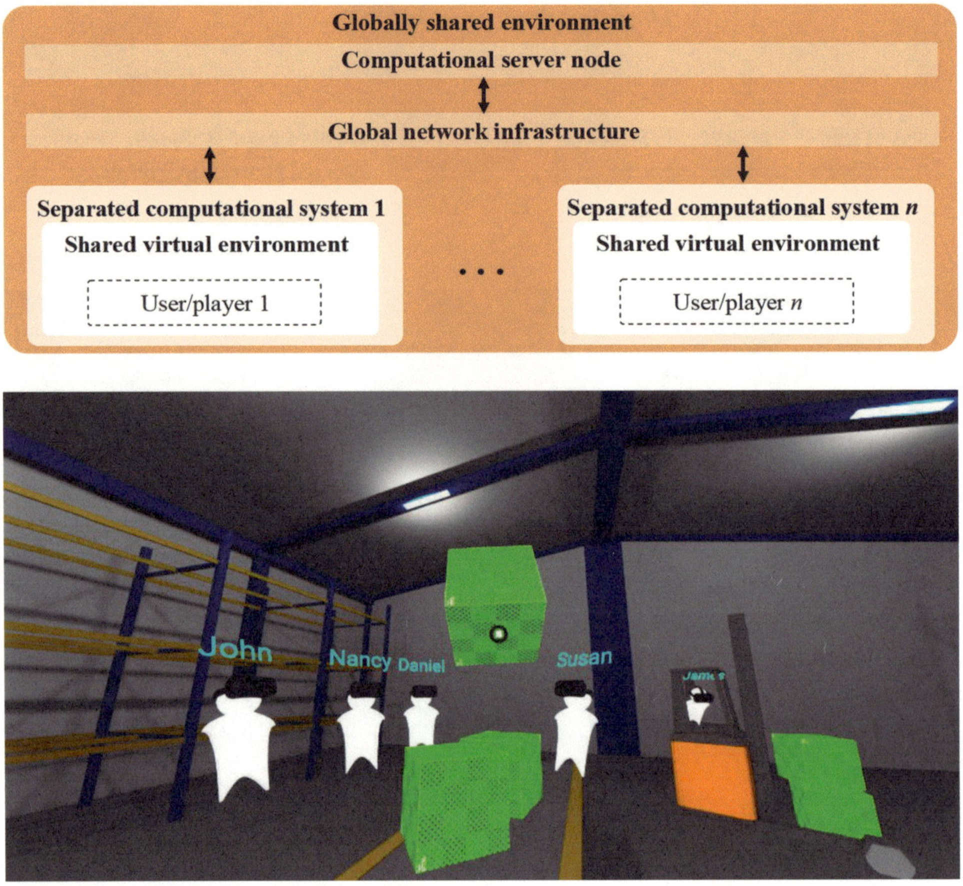

Figure 4.
Globally shared environment (LIRKIS DCI FEEI TU Košice).

which clearly needs to solve the availability of such environments for multiple and also different user devices and systems. As the development of new XR technologies grows rapidly, it is necessary to ensure the adaptability of systems and user interfaces so that they are able to handle new technological innovations. Optimization of data sharing is also most needed, as the problem of latency between users increases in the case of a global connection. Solving these problems can positively contribute to the expansion of virtual collaboration for any purpose.

3. Concepts of application architecture

An accurate application architecture concept is deployed for each of the user interaction sharing systems. The main task of this concept is to ensure the distribution of user data as well as an event sharing in collaborative virtual environments. In fact, different types of application architectures are deployed. The concept of these architectures depends on the network infrastructure and user connections.

Although each of the architectures is specific for its typical features, it is very rigorous to consider its advantages and disadvantages when using its concept. Therefore, the application architecture in systems of collaborative virtual environments is often adapted to the needs of the intended system. When creating multipurpose environments, it is possible to apply standard types of architectures without their components changing. However, extensions and modifications to application architectures are necessary for specific purpose environments.

3.1 Architectural model client: server

This architectural model uses the data centralization between users (clients) at the level of one common communication node (server). This type of application architecture is applicable to all types of collaborative virtual environments, and it is also popular in game applications.

The server (**Figure 5**) is responsible for obtaining and mediating each data/information submitted by the client and for their centralization. According to Doležal [26], the client–server architectural model use is required if targeted distribution of data among clients and management of their access is necessary. The main logic of virtual collaboration data sharing is, therefore, implemented primarily at the server level. It makes it possible to provide interaction data, solve different hierarchical levels of clients, operate possible data collection, and maintain a consistent collaboration [27]. Another advantage is the ease of clients' computing performance. Then, it is possible to use their available "power" just for the simulation of the collaborative virtual gaming environment.

The client–server architectural model use is fully dependent on the performance of the centralized node (server). Due to the increasing number of clients, higher demands on server performance are required. The server, therefore, mainly deals with data processing and replication among clients. To reduce the calculation requirements, it is possible to influence the interval of data sent by clients during the game or virtual collaboration [28]. Another way is to limit the number of active clients within the server performance capabilities.

If any of the specific subsystems are deployed on the client side, then the data from the subsystems are also replicated among the other clients. A subsystem may be represented by either a program unit or a device that is used additively and provides some of the whole system's additional functions (e.g., artificial intelligence and speech recognition).

The following subsystems can be considered as subsystems connecting to the client–server application architecture for the purpose of virtual collaboration:

- physical shared subsystem,
- the equipment operating on a sensor basis (sensor network),
- the application with the generation of data, events, and phenomena for the game environment.

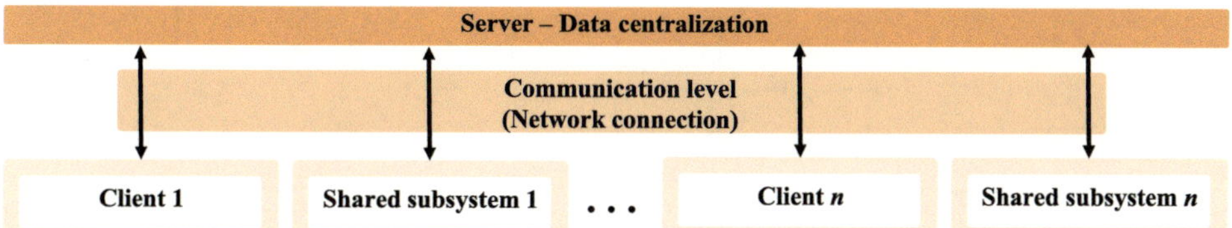

Figure 5.
Topology of virtual collaboration with the client–server architectural model.

Each of the subsystems can be represented by a specific module on the client side. Then, it is necessary to develop an optimal way of communication with other clients. Dias [29] highlights the potential for dynamically adaptive client–server architecture development. It can contain interfaces for obtaining data from various subsystems and its subsequent transformation for the virtual collaboration needs. The disadvantages of the mentioned architecture depend on the data centralization level. Increasing the number of interfaces at the server level can lead to frequent system-wide service outages or latency in the data processing. According to Khalid [30], these problems can be prevented by data size optimalization using compression.

3.2 Architectural model peer-to-peer

The second architectural model used to share virtual collaboration is the peer-to-peer model. The basic difference from the client–server model is the absence of a node (server), which centralizes the information provided by clients and their subsystems. Data sharing is replicated each other among clients (**Figure 6**). Each client actively sends its own data to all users in real time. The use of a peer-to-peer architecture is most suitable for data sharing only within smaller user groups [4]. Typical examples are environments/games with a limited number of users/players. This condition applies to both local and globally shared virtual environments.

When using the peer-to-peer model, the precondition for the emergence of requirements for higher performance of client devices is here, especially for the need to ensure communication with other clients [31]. In this way, it is complicated to optimize the client's performance so that an effective level of interaction as well as visual feedback can be achieved.

Data consistency is compromised when sharing user interactions if clients approach a collaborative environment with devices with different computing performances. In this case, there is a problem in synchronizing the multiuser interaction. Clients with higher computing performance are able to process interactions much faster compared to clients with low computing performance and high latencies. It is difficult to synchronize client interactions without losing any shared data [32], due to the absence of a central node (server). Another problem arises when globalization of such an architecture is needed. It is necessary to use a centralized communication node in global systems with a peer-to-peer architecture. Without its use, different latencies with high deviations can occur among clients.

This problem can manifest itself in a high inconsistency of shared data. This can be quite a problem, especially in action games. Due to these complications, the use

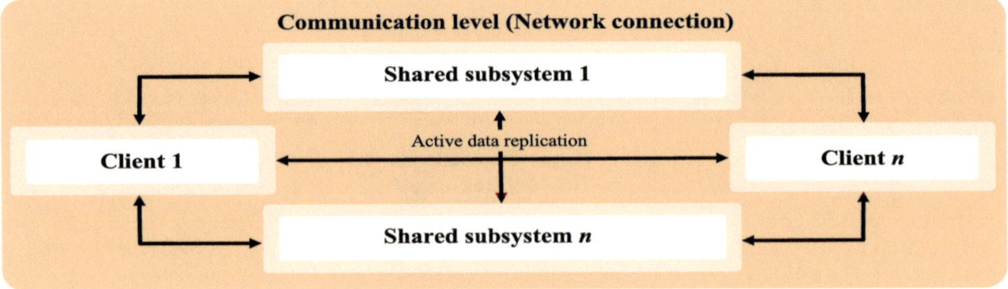

Figure 6.
Topology of virtual collaboration with the peer-to-peer architectural model.

of peer-to-peer architecture is only possible in purpose-specific environments with a limited number of users.

Despite the mentioned disadvantages, it is necessary to emphasize the principle of using this architecture as a suitable solution in the case of systems based on the client–server architectural model [33]. For the needs of failure of securing client connections in the event of client–server architecture failures, it is appropriate to use the peer-to-peer architecture as a backup mechanism for sharing game data or communication among players.

3.3 Distributed server architecture

There are cases when it is necessary to process complex databases or nonstandard data types. In the case of virtual collaboration, the deployment of virtual environments for specific purposes is often present. For example, in addition to the standard architecture, it is necessary to connect clients to other separate server nodes (e.g., to payment systems of games).

Simultaneous data transfer among multiple clients and servers is provided by the distributed server architecture, as shown in **Figure 7**. An example is the sharing of modeling CAD/editing systems [34]. In them, virtual collaboration depends on several externally available servers. Each of the servers is responsible for providing specific functionality. Then, it deploys full computing performance for this functionality. The main goal of the mentioned architecture is to separate the primary functionality associated with the virtual collaboration game and the secondary functionality providing specific data [35].

From a topological point of view, a distributed server architecture consists of at least one primary and one secondary server. The primary server manages clients' accesses, manages the sharing of user interactions, and ensures the availability of a common virtual/game environment. The secondary server is associated with different types of subsystems. These are used in addition and they need separate computing performance. The secondary server processes complex data needed for collaboration or game details.

During the collaboration, the data obtained from the server nodes are sorted and passed only to the specified context of the game environment [36]. The integrations of data from external servers (e.g., the weather conditions, the existing systems status, the notification, and payment data and records or the nodes with artificial intelligence for bots control) are frequent in simulation game systems.

However, if the primary and secondary servers were centralized in a single server, their level of reliability and services would be severely affected by computing

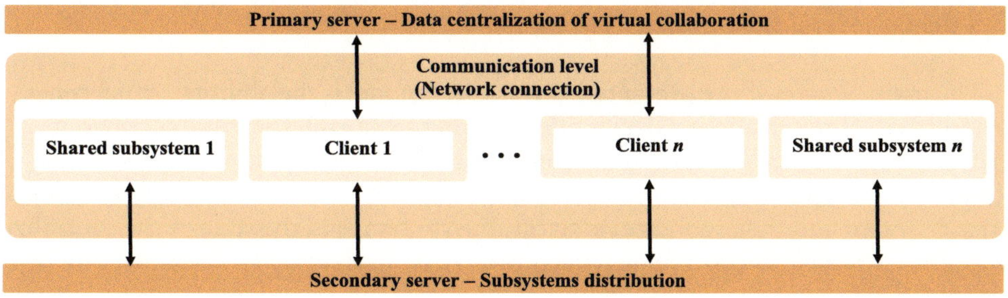

Figure 7.
Topology of using distributed server architecture in the concept of virtual collaboration.

performance decrease. This would also significantly limit its performance for complex computational operations.

4. Consistency models for sharing virtual game environments

To use specific application architectures, it is important to ensure the correct consistency of the virtual environment. In collaborative virtual environments, users/players perform various types of activities. These activities affect the modification of shared content, that is the virtual game environment. Ensuring consistency means that in case, any change is made to shared content by one user/player, this change will be visible to other users/players at the same time [37]. At the same time, the state of the virtual environment will be equivalent for all users/players after the change is made.

Consistency is directly related to the synchronization of virtual content between all users/players simultaneously. However, in this case, a problem arises if a group of users/players uses devices with different computing power as well as different network connection bandwidth. As stated in Ref. [38], there is a strong assumption that users with higher computing power and bandwidth could be prioritized when performing any interaction. This would strongly increase the number of conflicts between the activities of all users/players as well as misunderstandings in playing together.

To solve the mentioned issues, consistency models for sharing virtual environments are introduced, the main purpose of which is to manage and synchronize concurrency among user/player activities when working with virtual content.

4.1 Model of centralized virtual environment

Ensuring consistency of sharing through the centralized virtual environment model is primarily applicable to client–server application architectures (**Figure 8**). Every user/player accesses the centralized virtual environment through a network connection.

Consistency management is ensured by the following rules, as depicted in **Figure 8**:

1. Once connected, every user/player will receive a replication of the entire virtual environment, including its current state.

2. All connected users/players can perform object modification simultaneously.

3. After the modification, the centralized virtual environment is updated and the change is replicated to all users/players.

The main benefit of the presented consistency model is the ability to synchronize users/players through a centralized node, tracking all the modifications of shared objects in real time [39]. This mechanism is suitable for the needs of concurrent collaboration management, where it is necessary to quickly record the modification of virtual content and then replicate its status. Every change in the state of the centralized virtual environment is continuously recorded. After users/players disconnection, the last modified state of the virtual environment is preserved and is available for future collaboration [38]. From an implementation point of view, the centralized environment is easily customizable and distributable among all users/players in real time.

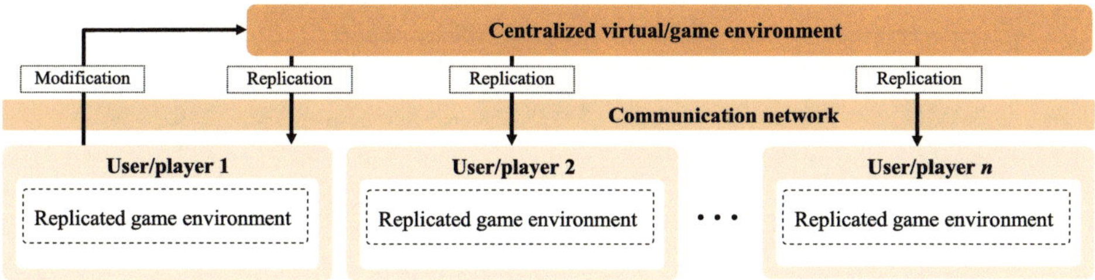

Figure 8.
Consistency model of centralized virtual environment sharing.

On the other hand, there are several principal limitations related to user access sharing and consistency management. A centralized virtual environment must be always available since it is the main source of virtual collaboration [40]. Although the service of this consistency model is easy to implement, it is important to optimize the centralized virtual environment for the needs of multiplatform access of users/players as well as XR technologies.

4.2 Model of actively replicated virtual environment

The data management consistency through active data replication among users/players is topologically identical to the peer-to-peer architecture. In the case of an actively replicated virtual game environment, consistency sharing is different because it is handled on the side of every user/player separately.

To maintain data consistency, it is necessary to implement a communication node (**Figure 9**). This is important mainly due to the need for global collaboration, as a result of which this model can be considered a hybrid with the client–server architecture [41]. Using the communication node, data consistency and user/player activity can be quickly sent and compared.

The model of an actively replicated virtual environment is controlled by the following algorithm, as depicted in **Figure 9**:

1. Before connecting, the shared virtual game environment must be programmatically present on every user device with the same initial state.

2. Once connected, only active data replication occurs between users/players via the communication node, not centralizing the data.

3. In case a new user/player joins the environment during the virtual collaboration, other users/players automatically replicate the current state of the virtual environment, with all the changes performed during the collaboration.

4. Modification of the virtual content can be performed by all users/players simultaneously.

5. If one of the users/players modifies some specific content, he replicates this event to the other users/players as a whole.

6. After completion of the collaboration, the latest state of the environment is not retained for future use.

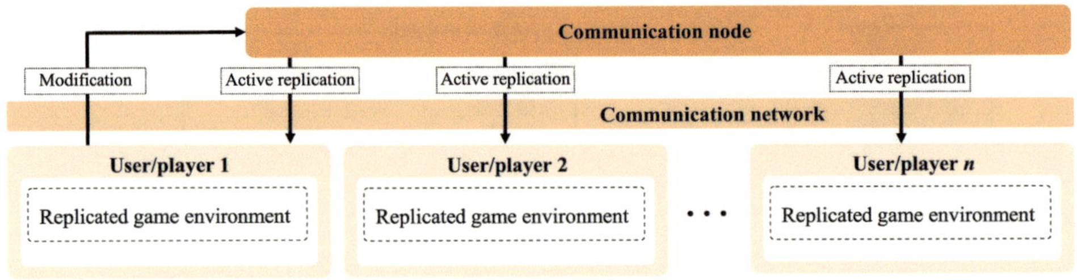

Figure 9.
Consistency model of active replication virtual environment sharing.

The benefits of this consistency model can be considered to be the high efficiency in synchronizing users/players, where the communication level is conducted directly, without processing by a third party [42]. This is desirable for some games. Despite the presence of the communication node, this is not data processing, but only replication to other users/players. Due to the close connection between users/players, it is possible to achieve a faster way of synchronizing and sharing collaborative activities.

Given the negatives, we can state that the growing number of users/players directly increases the requirements for the computing power of their devices. To extend the virtual collaboration systems with XR technology, it is necessary to emphasize that each of them has significantly different computing power. Therefore, the usability of active replication in sharing a virtual environment for a group of XR technologies is insufficient. Another resulting problem is the congestion of the communication level in case the given consistency model would be deployed for global use. In conclusion, this model can be considered effective only if the number of players does not change rapidly during the collaboration as well as the number of players is low.

4.3 Model of distributed virtual environment

In terms of data consistency, the sharing of distributed virtual environments is closely related to the model of centralized virtual environment sharing. From a topological point of view, the same client–server architecture is deployed (**Figure 10**). It is extended by an additional (secondary) centralized virtual environment.

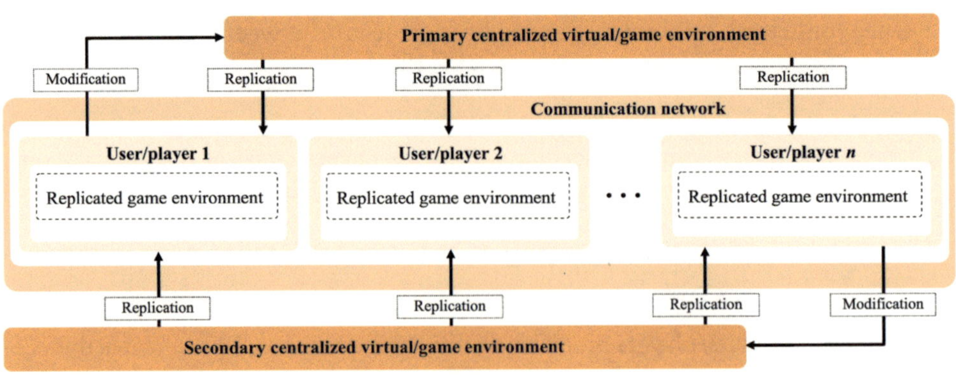

Figure 10.
Consistency model of distributed virtual environment sharing.

The primary centralized virtual/game environment, which provides the sharing of the main collaborative space, is considered to be the main one. The purpose of this space is to ensure the general existence of users/players and provide management of interactions and user/player entities.

Secondary environments provide an extension of virtual collaboration with specific virtual content. Their use would not be possible without primary environments since they are intended to broaden the context of collaboration. By combining the primary and secondary environments, a fully distributed collaborative virtual gaming space is created.

The algorithm of the data consistency management is identical to the algorithm of the centralized virtual environment model. Each of the users/players uses both the primary and the secondary environment, while the secondary environment work consists only of the content that is currently needed [43].

The main benefit of distributed virtual environments is the separation of general-purpose development from specifically defined functionality [44]. The same is true for mediating the virtual content and consistency at two levels of collaboration, which specifically controls user/player accesses and interactions. Due to the complexity of two interdependent environments in parallel, the application of such a consistency model is significantly challenging, mainly due to the problems of their joint synchronization [45]. Another negative is the need for duplicate development of system and user interfaces, especially if their intention is to support XR technologies.

5. Data ownership sharing

Data ownership sharing is the main mechanism for data consistency in managing multiuser interaction. Its aim is to ensure the transfer of virtual content among multiple users/players in real time [46]. Ownership of virtual content is applicable at the level of any application architecture. However, it can be effectively described, especially with the client–server architecture (**Figure 11**). The process of data ownership sharing consists of the following points:

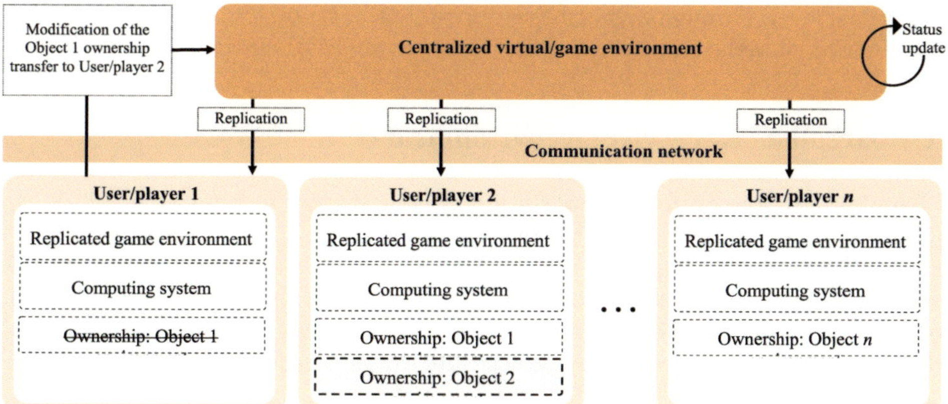

Figure 11.
Scheme of the data ownership sharing model.

- Ownership acquisition,
- Ownership transferring,
- Ownership leaving,
- Ownership enforcing.

The acquisition of ownership depends on the collaboration logic associated with the context of the shared virtual game environment. In general, when a user/player interacts with a shared virtual object, he becomes its owner. The same principle is used in the case of the creation of one's own virtual object by some of the users/players. When ownership is acquired, the centralized virtual environment is automatically modified and the change is replicated to other users/players. The following rule applies in virtual environment sharing: In concurrent collaboration, one user/player can own a group of objects, but one object cannot have more than one owner at a time [47].

The ownership transferring occurs when a user/player transfers his own object to one of the other users/players. In this case, there is a modification of the transfer of ownership to the centralized virtual environment and subsequently to a specific user/player. Other users/players of the collaborative group are also informed about the ownership transfer. These receive replication about the status of the ownership transfer to a particular user/player.

The ownership leaving occurs when one of the users/players relinquishes the virtual object and he leaves it in a centralized virtual environment. Subsequently, the availability of the object is replicated to all users/players. Then, the object is available to players and it can be acquired.

Data ownership sharing can be also affected by specific collaboration or game rules. An example is the division of users/players into individual hierarchical levels. Then, the whole strategy of ownership acquiring and transferring changes according to these levels. The examples are the systems for training or instruction in a gamified form. In them, the users/players are most often divided into groups of editors, instructors, collaborators, or observers. In such systems, there are possibilities for ownership enforcement. This ability is usually available to users/players at the highest hierarchical level. Through ownership enforcement, it is possible to acquire a virtual object without it being transferred or left by its owner in advance.

6. Global collaborative virtual environment (LIRKIS G-CVE)

Based on the above information, the global collaborative web system LIRKIS G-CVE [48] was implemented as the main working platform for these systems in the LIRKIS laboratory (**Figure 12**). The system was created as a result of the web XR system architectural model formalization. The purpose of the system is to provide a globally accessible virtual environment through web browsers. LIRKIS G-CVE uses a client–server architecture to provide access to the shared environment to multiple clients in real time.

The main part of the system is a remote web server. This manages communication among clients and mediates their interactions in a shared virtual environment [49]. The server uses three application software frameworks: Node.js, Express.js, and Networked-Aframe implemented in the JavaScript scripting language. These

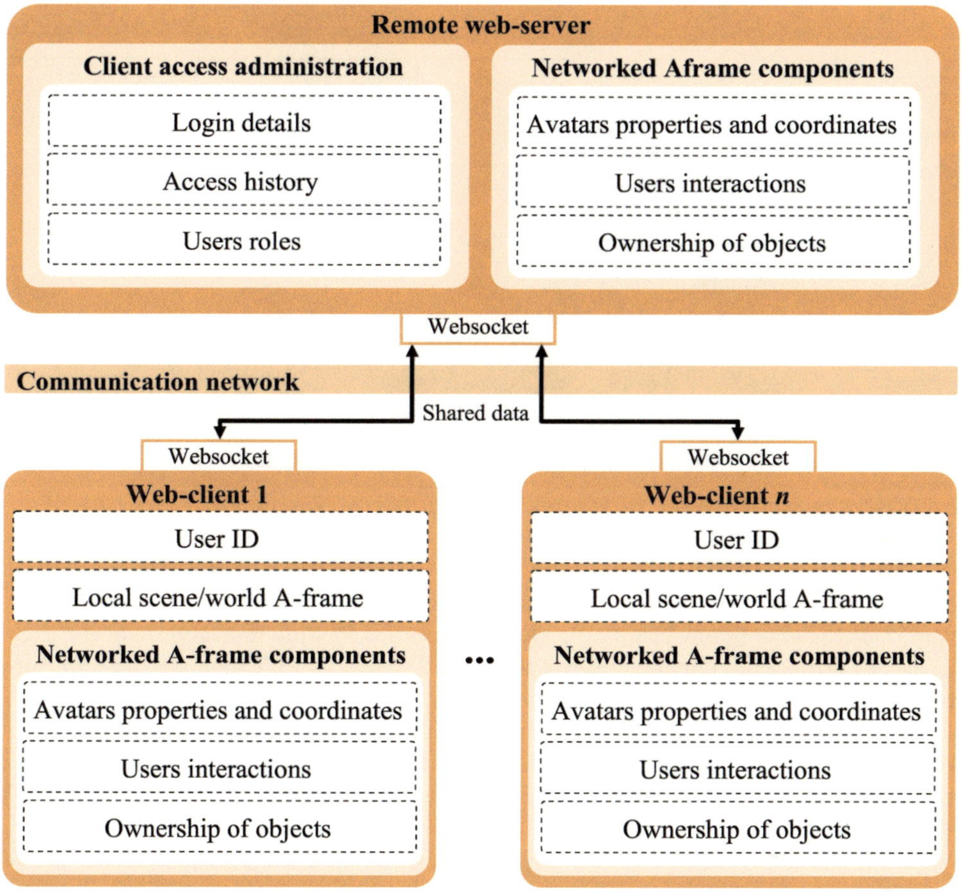

Figure 12.
LIRKIS G-CVE architecture.

frameworks ensure the execution of services in the system application background. Node.js provides real-time parallel client connections and it manages asynchronous client–server data communication. Express.js provides server rest APIs to handle HTTP requests from clients during a connection. Networked-Aframe manages the sharing and synchronization of the virtual environment among clients, including all virtual 3D objects and interactions.

The second part presents the client's web interface, which is displayed through a web browser and performs asynchronous rendering of the shared virtual environment. *A-Frame* technology is used in the implementation of this interface with regard to the guaranteed support of several platforms and the development of interactive virtual environments. *A-Frame* is based on the Three.js visualization library, which provides an application interface for creating 3D computer graphics in a web browser. Using *A-Frame*, virtual environments are created at the HTML level and their functionality is implemented in JavaScript. Using asynchronous JavaScript (AJAX), it is possible to render all changes in the environment simultaneously without the need to reload the entire interface.

Communication between client and server is based on the *WebSocket* protocol. Each collaborative virtual environment is accessible to clients under a specific URL. After the client connects, the server requests his data and then provides it to other

Figure 13.
Example of BUS simulator virtual environments.

active clients. The amount of data transferred among clients depends on the characteristics of the shared environment. These properties include interaction levels, the number of connected clients, and the amount of data shared. Client–server communication depends on the width and speed of the network connection. This is why in case of long-term inactivity, the inactive clients are automatically disconnected. This prevents the server from being overloaded by inactive network entities. If the client's connection is terminated, the server removes its virtual entity (including the avatar) from the environment and closes the connection.

Part of the implementations in the previous works and projects of the LIRKIS laboratory was the creation of virtual collaborative environments for the purpose of

Figure 14.
Example of gamified training–testing application on safety in the construction industry.

Figure 15.
Example of a virtual drone game.

Figure 16.
Example of virtual DCI OpenLAB environments.

a global collaboration of users in gamified form as well. Each of the training games supports a cross-platform approach and user interaction. As an example, we present a bus simulator (**Figure 13**), a gamified training–testing application on safety in the construction industry (**Figure 14**), a collaborative environment for training work with drones in a playful way (for artificial intelligence training data) (**Figure 15**), and an environment for testing multiuser interaction (department virtual OpenLAB environment, DCI FEEI TU Košice) (**Figure 16**).

7. Conclusion

The computer games industry and research are progressing quite well, although it requires significant financial resources. As the demands of users/players also increase, it is good if this technology is a very immersive experience for its users/players. The games allow to bring also education, training, or presentation of various kinds of designs up to an entirely new level. The XR technology usage is one of the ways in which this can be achieved. It represents a new form of visualization, a new form of experiences, a new form of feeling, and a new form of socially shared life. Models can be created using 3D modeling tools, including CAD software, or by using 3D scanners or motion capture systems and inserted into, for example, a real scene. A new game era is starting.

Acknowledgements

This work was supported by the Operational Programme Integrated Infrastructure in the frame of the project: Intelligent systems for UAV real-time operation and data processing, code ITMS2014+: 313011 V422, and co-financed by the European Regional Development Fund and by Slovak KEGA Agency under Grant no. "002TUKE-4/2021 Implementation of Modern Methods and Education Forms in the Area of Cybersecurity towards Requirements of Labour Market."

Author details

Branislav Sobota[1*], Marián Hudák[2] and Emília Pietriková[1]

1 Technical University of Košice, Slovakia

2 Siemens Healthcare s.r.o. Košice, Slovakia

*Address all correspondence to: branislav.sobota@tuke.sk

References

[1] Margetis G, Papagiannakis G, Stephanidis C. Realistic natural interaction with virtual statues in x-reality environments. In: International Archives of the Photogrammetry, Remote Sensing & Spatial Information Sciences. 2019

[2] Fast-Berglund Å, Gong L, Li D. Testing and validating extended reality (xr) technologies in manufacturing. Procedia Manufacturing. 2018;**25**:31-38

[3] Huzaifa M, Desai R, Jiang X, Ravichandran J, Sinclair F, Adve SV. Exploring extended reality with ILLIXR: A new playground for architecture research. 2020. arXiv: abs/2004.04643 n.pag

[4] Suslov N. Livecoding. space: Towards p2p collaborative live programming environment for webxr. In: Proceedings of the Fourth International Conference on Live Coding (ICLC 2019), Medialab Prado; Madrid, Spain. 2019

[5] Seo D, Yoo B, Ko H. Webizing collaborative interaction space for cross reality with various human interface devices. In: Proceedings of the 23rd International ACM Conference on 3D Web Technology. 2018. pp. 1-8

[6] Anthes C, García-Hernández RJ, Wiedemann M, Kranzlmüller D. State of the art of virtual reality technology. In: 2016 IEEE Aerospace Conference. 2016. pp. 1-19

[7] Wang S-M, Wang Y-C, Pan W-J, Lin C-Y. Developing innovative cross-platform web-based virtual reality services. In: 2020 IEEE International Conference on Consumer Electronics-Taiwan (ICCE-Taiwan). 2020. pp. 1-2

[8] Paiva PV, Machado LDS, Valença AMG, De Moraes RM, Batista TV. Enhancing collaboration on a cloud-based CVE for supporting surgical education. In: 2016 XVIII Symposium on Virtual and Augmented Reality (SVR). 2016. pp. 29-36

[9] Hudák M, Korečko Š, Sobota B. Lirkis global collaborative virtual environments: Current state and utilization perspective. Open Computer Science. 2020;**11**(1):99-106

[10] Nguyen TTH, Duval T. A survey of communication and awareness in collaborative virtual environments. In: 2014 International Workshop on Collaborative Virtual Environments (3DCVE). 2014. pp. 1-8

[11] Hudák M, Korečko Š, Sobota B. Special input devices integration to lirkis cave. Open Computer Science. 2018;**8**(1):1-9

[12] de Back TT, Tinga AM, Nguyen P, Louwerse MM. Benefits of immersive collaborative learning in cave-based virtual reality. International Journal of Educational Technology in Higher Education. 2020;**17**(1):1-18

[13] Hudák M, Korečko Š, Sobota B. On architecture and performance of lirkis cave system. In: 2017 8th IEEE International Conference on Cognitive Infocommunications (CogInfoCom). 2017. pp. 000295-000300

[14] Hudák M, Korečko Š, Sobota B. Peripheral devices support for lirkis cave. In: 2017 IEEE 14th International Scientific Conference on Informatics. 2017. pp. 117-121

[15] Hudák M. Extension of web-based collaborative virtual environments for mixed reality interfaces. In: SCYR 2020. 2020. pp. 97-98

[16] Dey A, Piumsomboon T, Lee Y, Billinghurst M. Effects of sharing physiological states of players in a collaborative virtual reality gameplay. In: Proceedings of the 2017 CHI Conference on Human Factors in Computing Systems. 2017. pp. 4045-4056

[17] Lee B, Hu X, Cordeil M, Prouzeau A, Jenny B, Dwyer T. Shared surfaces and spaces: Collaborative data visualisation in a co-located immersive environment. In: IEEE Transactions on Visualization and Computer Graphics. 2020

[18] Horst R, Alberternst S, Sutter J, Slusallek P, Kloos U, Dörner R. Avatar2avatar: Augmenting the mutual visual communication between co-located real and virtual environments. In: VISIGRAPP (2: HUCAPP). 2019. pp. 89-96

[19] Lacoche J, Pallamin N, Boggini T, Royan J. Collaborators awareness for user cohabitation in co-located collaborative virtual environments. In: Proceedings of the 23rd ACM Symposium on Virtual Reality Software and Technology. 2017. pp. 1-9

[20] Šašinka Č, Stachoň Z, Sedlák M, Chmelík J, Herman L, Kubíček P, et al. Collaborative immersive virtual environments for education in geography. ISPRS International Journal of Geo-Information. 2019;**8**(1):3

[21] Piumsomboon T, Day A, Ens B, Lee Y, Lee G, Billinghurst M. Exploring enhancements for remote mixed reality collaboration. In: SIGGRAPH Asia 2017 Mobile Graphics & Interactive Applications. 2017. pp. 1-5

[22] Du J, Shi Y, Zou Z, Zhao D. Covr: Cloud-based multiuser virtual reality headset system for project communication of remote users. Journal of Construction Engineering and Management. 2018;**144**(2):04017109

[23] Alatalo T, Koskela T, Pouke M, Alavesa P, Ojala T. Virtualoulu: Collaborative, immersive and extensible 3d city model on the web. In: Proceedings of the 21st International Conference on Web3D Technology. 2016. pp. 95-103

[24] Wu T-H, Wu F, Liang C-J, Li Y-F, Tseng C-M, Kang S-C. A virtual reality tool for training in global engineering collaboration. Universal Access in the Information Society. 2019;**18**(2):243-255

[25] El-Kassas WS, Abdullah BA, Yousef AH, Wahba AM. Taxonomy of cross-platform mobile applications development approaches. Ain Shams Engineering Journal. 2017;**8**(2):163-190

[26] Dolezal M, Chmelik J, Liarokapis F. An immersive virtual environment for collaborative geovisualization. 2020. p. 4. arXiv: abs/2010.06279

[27] García AS, Fernando T, Roberts DJ, Bar C, Cencetti M, Engelke W, et al. Collaborative virtual reality platform for visualizing space data and mission planning. Multimedia Tools and Applications. 2019;**78**(23):33191-33220

[28] Hudák M, Sivý M, Sobota B. Uniform Smartphone Controller for Web-Based Virtual Reality Purposes. Košice (Slovensko): Fakulta elektrotechniky a informatiky; 2021

[29] Dias DR, Guimarães MP, Kuhlen TW, Trevelin LC. A dynamic-adaptive architecture for 3d collaborative virtual environments based on graphic clusters. In: Proceedings of the 30th Annual ACM Symposium on Applied Computing. 2015. pp. 480-487

[30] Khalid S, Ullah S, Alam A, Din F. Optimal latency in collaborative virtual environment to increase user performance: A survey. International Journal of Computer Applications. 2016;**142**(3):35-47

[31] Buyukkaya E, Abdallah M, Simon G. A survey of peer-to-peer overlay approaches for networked virtual environments. Peer-to-peer Networking and Applications. 2015;**8**(2):276-300

[32] Capece N, Erra U, Losasso G. Design and implementation of a web-based collaborative authoring tool for the virtual reality; proc. of international conference "15th International Conference on Signal-Image Technology & Internet-Based Systems (SITIS)". Sorrento, Naples, Italy; 2019. pp. 603-610

[33] Desprat C, Jessel J-P, Luga H. 3devent: A framework using eventsourcing approach for 3d web-based collaborative design in p2p. In: Proceedings of the 21st International Conference on Web3D Technology. 2016. pp. 73-76

[34] Okuya Y, Ladeveze N, Gladin O, Fleury C, Bourdot P. Distributed architecture for remote collaborative modification of parametric cad data. In: 2018 IEEE Fourth VR International Workshop on Collaborative Virtual Environments (3DCVE). 2018. pp. 1-4

[35] Elfizar E, Baba MS, Herawan T. Object-based viewpoint for large-scale distributed virtual environment. Malaysian Journal of Computer Science. 2015;**28**(4):301-317

[36] Shen S, Hu S-Y, Iosup A, Epema D. Area of simulation: Mechanism and architecture for multi-avatar virtual environments. ACM Transactions on Multimedia Computing, Communications, and Applications (TOMM). 2015;**12**(1):1-24

[37] Roberts D, Wolff R. Controlling consistency within collaborative virtual environments. In: Eighth IEEE International Symposium on Distributed Simulation and Real-Time Applications. 2004. pp. 46-52

[38] Fleury C, Duval T, Gouranton V, Arnaldi B. Architectures and mechanisms to efficiently maintain consistency in collaborative virtual environments. In: Proceedings of Software Engineering and Architectures for Realtime Interactive Systems-SEARIS. 2010a. pp. 87-94

[39] Yarsquo A, Ismail AS, Subramaniam S, et al. On consistency and security issues in collaborative virtual environment systems. International Journal of Physical Sciences. 2013;**8**(33):1646-1654

[40] Gital AY, Ismail AS, Chiroma H, Abubakar AI, Abdulhamid BM, Maitama IZ, et al. Performance analysis of cloud-based cve communication architecture in comparison with the traditional client server, p2p and hybrid models. In: The 5th International Conference on Information and Communication Technology for The Muslim World (ICT4M). 2014. pp. 1-6

[41] Fleury C, Duval TH, Gouranton V, Arnaldi B. A new adaptive data distribution model for consistency maintenance in collaborative virtual environments.; In: JVRC 2010 (2010 Joint Virtual Reality Conference of EuroVR - EGVE - VEC). 2010. DOI: 10.2312/EGVE/JVRC10/029-036

[42] Paiva PV, Machado L, de Oliveira JC, de Moraes RM. Networking issues for 3d medical collaborative virtual environments: Design and applications. SBC Journal on Interactive Systems. 2017;**8**(2):89-97

[43] Millar JR, Hodson DD, Peterson GL, Ahner DK. Consistency and fairness in real-time distributed virtual environments: Paradigms and relationships. Journal of Simulation. 2017;**11**(3):295-302

[44] Yusen L, Deng Y, Cai W, Tang X. Fairness-aware update schedules for improving consistency in multi-server distributed virtual environments. In: Proceedings of the 9th EAI International Conference on Simulation Tools and Techniques. 2016. pp. 1-8

[45] Zhang W, Zhou H-J. A fairness oriented consistency control method in distributed virtual environment systems. In: DEStech Transactions on Computer Science and Engineering (CST). 2017

[46] Zhou M, Leenders MA, Cong LM. Ownership in the virtual world and the implications for long-term user innovation success. Technovation. 2018;**78**:56-65

[47] Lange P, Weller R, Zachmann G. Scalable concurrency control for massively collaborative virtual environments. In: Proceedings of the 7th ACM International Workshop on Massively Multiuser Virtual Environments. 2015. pp. 7-12

[48] Hudák M, Korečko Š, Sobota B. Advanced user interaction for web-based collaborative virtual reality. In: 2020 11th IEEE International Conference on Cognitive Infocommunications (CogInfoCom). 2020. pp. 000343-000348

[49] Hudák M, Korečko Š, Sobota B. Enhancing team interaction and cross-platform access in web-based collaborative virtual environments. In: 2019 IEEE 15th International Scientific Conference on Informatics. 2019. pp. 000171-000176

Chapter 2

Commercial-off-the-Shelf (COTS) Games: Exploring the Applications of Games for Instruction and Assessment

Diana R. Sanchez, Amanda Rueda, Leila Jimeno Jimènez and Mahsa Norouzi Nargesi

Abstract

Despite the growing interest in utilizing commercial off-the-shelf (COTS) games for instructional and assessment purposes there is a lack of research evidence regarding COTS games for these applications. This chapter considers the application of COTS games for instruction and assessment and provides preliminary evidence comparing COTS game scores to traditional multiple-choice assessments. In a series of four studies, we collected data and compared results from the performance in a COTS game to scores on a traditional multiple-choice assessment written for the purposes of each study. Each assessment was written to evaluate the same content presented in the game for each respective study. Three of the four studies demonstrated a significant correlation between the COTS game and the traditional multiple choice assessment scores. The non-significant value in Study 4 was likely due to a small sample size ($n < 100$). The results of these studies support our hypothesis and demonstrate that COTS games may be a useful educational tool for training or assessment purposes. We recommend that future research focuses on specific applications of COTS games to explore further opportunities for utilizing COTS in education and assessment.

Keywords: commercial off-the-shelf (COTS) game, game-based assessments (GBAs), serious games, game-based learning (GBL), instructional games, multiple-choice assessments

1. Introduction

Over the last decade, the use of games for purposes outside of entertainment has grown in both research and industry, leading to discrepancies in conceptual definitions, use cases, and evidence-based benefits [1–4]. An increase in recent research literature reviews and meta-analyses on the topic of game-based interventions emphasizes the wealth of research being conducted in this area [5, 6]. However, game-developers have quickly capitalized on the widespread interest and often market game-based interventions in ways that are not always empirically tested with evidence for their publicized use [7]. In this chapter we aim to provide clarity on this topic by first reviewing important terminology, second discussing applications of games for instruction, and third exploring applications of games for assessment. Then we focus on the rise of commercial off-the-shelf (COTS) games for both instructional and assessment purposes. We close by highlighting the authors' research from four studies focused on comparing COTS games to traditional assessments.

1.1 Important terminology

In this chapter, we use the term *game-based intervention* to speak broadly of game-like experiences and the ways they are applied for non-entertainment purposes such as education or evaluation. Within this broad conceptualization, there are several important terms: game-based learning (GBL), game-based assessment (GBA), gameful design, gamification, and COTS games. GBL, also referred to as serious games [8] occurs when a game is used for the primary purpose of learning [2]. Similarly, GBAs are evaluations that take place within the context of a game expressly designed to identify specific skills, traits, or behaviors [9]. In contrast, gameful design and gamification are design strategies which incorporate game elements into new or existing assessments respectively; typically for the purpose of engaging and motivating individuals [2, 9]. Thus, a thorough understanding of game design is requisite to the development of effective GBLs and GBAs as these interventions use the full suite of game attributes. For example, see [1] for a taxonomy of game attributes. As a result, developing GBL and GBAs entails considerably more expertise and resources compared to incorporating game elements through gamification or gameful design. A potential solution to the resource-heavy nature of game development is the utilization of COTS games. Designed by game developers, COTS are most commonly designed for entertainment purposes and are available for purchase by a vendor and can be used almost immediately after purchase [10, 11]. COTS games offer a reasonable alternative to the time and resource intensive commitment to develop one's own games [12, 13]. GBL, GBA, and COTS games are explored in more detail in the following sections.

1.2 Game-based learning (GBL)

Games have been used to stimulate student learning by attracting students' attention through increased motivation [14]. Additionally, researchers and practitioners have shown increasing interest in alternative instructional methods, which emphasize a more active learner role [8]. Combined with the increasing growth of interactive technologies, this provides a unique opportunity to explore learning environments which involve students in their own learning process, such as the gamification of learning content and using a GBL solution [15]. In contrast with traditional learning methods, educational games have shown a positive relationship with the thinking skills of learners and can improve learner motivation [16, 17]. Considering the ways that GBL solutions can present learning experiences in a motivating way to students, using games for learning could feasibly increase student engagement and in turn have the potential to improve the academic performance of students [18–20].

1.3 Game-based assessment (GBA)

GBAs are designed to measure knowledge, skills, or abilities (KSAs) within the context of the game [21–23]. They are defined as evaluations that use game elements to immerse the individual in a specific game environment, allowing them to interact with it, and demonstrate the desired KSA [2]. Importantly, GBAs typically use game activities as tasks to generate evidence of complex skills [24]. Using a GBA with strong psychometric evidence (e.g., reliability and validity evidence for the intended use) has the potential to provide a number of benefits. Some researchers have proposed that GBAs might be designed to generate positive outcomes in test-takers such as reduced test anxiety, challenges faking, and measuring more behavior-based measures [3, 7–9, 16, 25, 26]. In the context of work, some researchers have referred to GBA as "an assessment method in which job candidates are players participating in a core gameplay loop while trait information is inferred" [9].

While traditional assessments use participants' responses to textual or graphic prompts to collect data about their knowledge, skills, and abilities, GBAs gather this data based on the test-taker's in-game behaviors [27]. These behaviors can be a range of information such as overt information like a choice a player makes when faced with a discrete decision in a game. A theory-driven assessment often focuses on a theoretical research model to build the design and intention of the evaluative components into the assessment. These components may lead to a narrative or series of decisions around targeted behaviors related to the variable being assessed. An alternative to a theory-driven approach are games built using a data-driven approach where minor behaviors during the gameplay are used to predict the outcomes of interest. A data-driven GBA is not built with a targeted construct and measurement in mind [28–30] but can empirically measure a construct using *trace data* such as mouse clicks, movements, interactions with objects in games, or time spent on a task [28]. This *trace data* can be collected automatically by a game and has been shown to demonstrate meaningful information with regards to assessing player information [28]. Either type of GBA changes how assessments are traditionally measured, but holds onto the psychometric properties within the game to evaluate a variety of KSAs [9, 31].

There are several benefits regarding GBAs that make them particularly appealing to organizations who might use them for personnel selection. One of these is the suggestion that GBAs might be able to predict job performance beyond traditional selection methods [16, 32]. Given the longstanding concern of applicant faking, the potential to reduce socially desirable responses are another benefit that GBAs may provide to organizational assessments [16, 32]. Since GBAs can be designed to measure traits and behaviors indirectly, this has the potential to obscure the purpose of the assessment which could in turn make it more challenging for candidates to identify the variable being assessed and what a good answer would look like [32]. A last benefit that may be of particular interest to organizations is that GBAs have been shown to reduce adverse impact compared with traditional paper-and-pencil tests [27, 33, 34]. Because of these benefits, organizations may want to consider investing in GBAs to evaluate different characteristics in the workplace, such as cognitive ability, individual characteristics, or job skills [27, 35].

1.4 Commercial off-the-shelf (COTS) games

Given the time, money, and expertise needed to develop game-based interventions, there are contexts in which it is more reasonable for researchers or practitioners to use an existing COTS game for their purpose rather than investing the time or resources into developing their own game [3, 27, 36]. However, a critical consideration when choosing to use a COTS game is that these games are rarely designed for the purpose they will be used for. Although a growing number of vendors are developing COTS games for learning or assessment purposes and having a growing body of evidence to support those games for those uses, in most contexts a researcher and practitioner will not have the option of a COTS game designed for the purpose they intend. This makes the consideration and selection of the COTS game to be used a critical step.

If a COTS game is being used for assessment or evaluation, the behaviors being displayed by the player must be measurable, either by a metric captured in the game or by an observable behavior that can be recorded by an observer. Additionally, the content needs to present a scenario in which the player has the opportunity to demonstrate relevant behavior to the variable being measured. For example, see the series of studies by [27] where VR COTS games were used for measuring spatial recognition variables.

If a COTS game were being used for a learning application, the game would need to present the relevant content that the player needs to learn. This includes presenting the information and ideally presenting a context in which the relevant information could be used or the intended skill could be practiced. It would be further valuable for the selection of the COTS game to avoid games with excessive information or components that aren't relevant to the learning experience, also called *seductive details* [37].

Research on traditional assessments focuses heavily on the psychometric reliability and validity of the instrument [9], but empirical support for the psychometric properties of COTS games is still in its infancy. This lack of research of the efficacy and utility of COTS games for instructional and assessment purposes has led to the current series of studies which are intended to provide preliminary evidence on the use of COTS game scores. In these studies, we ask the following research question.

Research question: *Will scores on COTS games significantly converge with traditional multiple-choice assessment scores?*

Below we detail four correlational studies focused on the application of COTS games. We seek to better understand potential applications of COTS games as learning or assessments interventions by better understanding the similarity between scores produced in a COTS game to traditional multiple-choice assessment. Each of the COTS games in the studies below was carefully considered and chosen as it represents content that could be relevant to learners in a GBL context. Since the scores produced by COTS games are not intended for learning or assessment purposes, our question is general to understanding if these scores are meaningfully related to a relevant measure in these particular cases; scores on a traditional multiple-choice assessment in the same content area. We caveat that this evidence does not generalize to all COTS games and individual data would need to be collected and analyzed if a COTS game is being considered for a similar application. However, this series of studies does provide a general proof-of-concept that may aid future applications of COTS games.

2. Research series

Data from the first three studies were collected as part of other projects and analyzed for this research question. All four studies collected samples from undergraduate and graduate students from Universities in the Western United States, with approval from the associated Institutional Review Board. Each study reports demographic information, scores from the COTS game used, and scores from the multiple-choice assessment. An image of the COTS game used in each study is shown in **Figure 1**. Each multiple-choice assessment included four response options, one

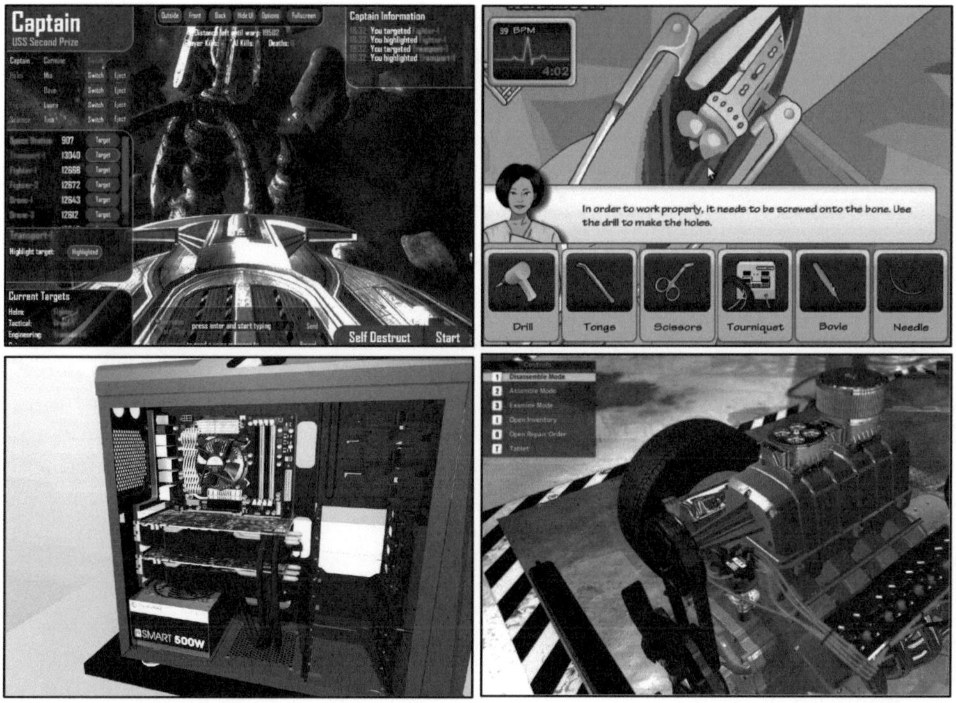

Figure 1.
Images from the commercial-off-the-shelf (COTS) games used in Study 1 (Quintet, top left), Study 2 (Arm Surgery 2, top right), Study 3 (PC Building Simulator, bottom left), and Study 4 (Car Mechanic Simulator, bottom right).

correct answer, and good metrics based on a pilot sample (i.e., reasonable difficulty between 0.30 and 0.90 and knowledge discrimination of $r > 0.25$). Each assessment was developed using research-based principles [37, 38]. Lastly, each study collected the participant's tendency to play videogames using the Video Game Pursuit scale (VGPu) [39]. The scores on this 19-item measure are reported on a 5-point scale from 1 = strongly disagree to 5 = strongly agree. This measure is used since researchers have proposed that a propensity towards playing video games may impact the results of videogame interventions [40].

2.1 Study 1

Participants were 385 students; primarily female (51%) and Caucasian 73% with an average age of almost 19 years (SD = 1.72). In the study, participants completed a consent form, then an initial survey measuring demographic information and video game pursuit. Participants then played the assigned game for 20 minutes followed by a second questionnaire which included a multiple-choice assessment. The study concluded with a four-minute debriefing video.

2.1.1 COTS game

In the videogame *Quintet*, participants play as crew members of a spaceship (i.e., captain, helm, tactical, engineer, or scientist) who must work with a team to complete various science missions. Participants must learn their role and how to manage their responsibilities on the ship to earn points and meet the mission objectives.

2.1.2 Multiple-choice assessment

A 26-item multiple choice assessment was written for this study with questions linked to the specific missions, roles, and responsibilities in the game.

2.2 Study 2

Participants were 140 students; primarily female (69%), and mostly Caucasian (39%) or Hispanic (29%) with an average age of almost 24 years ($SD = 4.85$). In the study, participants completed a consent form, then played the assigned game for 15 minutes followed by a questionnaire with the multiple-choice assessment, demographic questions, and the measure of videogame pursuit. The study concluded with a short debriefing statement.

2.2.1 COTS game

The videogame *Arm Surgery 2*, takes place in a virtual operating room where players take on the role of a surgical doctor. The game begins with a tutorial, where a nurse guides the player through surgical techniques and medical instruments. Players then have to complete a surgery to repair a broken arm as quickly as possible with as few mistakes as possible.

2.2.2 Multiple-choice assessment

A 17-item multiple choice assessment was written for this study. All questions pertained to the surgical terms, tools, and procedures presented during the game.

2.3 Study 3

Participants were 100 students; mostly females (67%) of Caucasian (40%) or Hispanic (32%) descent with an average age of about 23 years ($SD = 6.28$). In the study, participants signed a consent form then reviewed an 8-page tutorial on how to play the game. Participants then played the assigned game for 30 minutes before completing a survey of demographic questions, videogame pursuit, and a multiple-choice assessment. Participants were then given a debriefing handout and dismissed.

2.3.1 COTS game

In the game *PC Building Simulator*, players take on the role of a PC repair shop owner. They must manage their shop by diagnosing and repairing computers to make money for their store.

2.3.2 Multiple-choice assessment

An 18-item multiple choice assessment was written for this study. All questions were related to the tasks and terminology used in the game for repairing PCs.

2.4 Study 4

Participants were 78 students; mostly females (70%) of Hispanic (27%) or Asian (28%) descent with an average age of about 23 years (SD = 6.28). In the study, participants completed a consent form then played the assigned game for 15 minutes. The study ended with a survey of demographic measures, videogame pursuit, and a multiple-choice assessment.

2.4.1 COTS game

In the game *Car Mechanic Simulator 2*, players take on the role of a mechanic to repair, paint, and tune cars in a garage. Their goal is to make money by repairing as many cars as possible.

2.4.2 Multiple-choice assessment

A 10-item multiple choice assessment was written for this study. All questions asked about the tools and types of repairs done in the game.

3. Research results

Results of the correlation coefficient analyses are shown in **Table 1**. These demonstrate that there are positive, significant correlation between the COTS game and multiple-choice assessments in three of the four studies; Study 1 $r = 0.23, p < 0.01$; Study 2 $r = 0.27, p < 0.01$; and Study 3 $r = 0.36, p < 0.01$; but not Study 4 $r = 0.02$, $p < 0.05$. This shows that in three of the four studies presented here, the score generated by playing the COTS game was significantly related to the score earned by the participant on the traditional multiple-choice assessment. These results are valuable for the area of COTS games because they show that when selected carefully, the scores produced in a COTS game might provide meaningful information for individuals wanting to use COTS games for learning or assessment purposes.

	M	*SD*	1	2	3	4	5
Study 1							
1. Sex at birth	0.48	0.50	—				
2. Age	18.97	1.72	0.18**	—			
3. Video game pursuit	2.64	0.99	0.62**	0.10*	**0.94**		
4. COTS game	0.55	0.15	0.16**	0.04	0.14**	—	
5. Multiple-choice assessment	0.56	0.15	0.29**	0.07	0.45**	0.23**	—
Study 2							
1. Sex at birth	0.28	0.44	—				
2. Age	23.75	4.85	0.14	—			
3. Video game pursuit	2.16	1.01	0.60**	−0.12	**0.91**		
4. COTS game	0.60	0.24	0.05	−0.26**	0.21*	—	
5. Multiple-choice assessment	0.11	0.03	0.05	0.16	−0.01	0.27**	—

	M	*SD*	1	2	3	4	5
Study 3							
1. Sex at birth	0.29	0.46	—				
2. Age	22.49	5.61	0.15	—			
3. Video game pursuit	2.90	0.55	0.59**	−0.13	**0.91**		
4. COTS game	0.55	0.27	0.11	−0.16	0.22*	—	
5. Multiple-choice assessment	0.71	0.16	0.09	−0.04	0.23*	0.36**	—
Study 4							
1. Sex at birth	0.20	0.40	—				
2. Age	24.04	6.85	0.29*	—			
3. Video game pursuit	2.51	1.10	0.50**	0.10	**0.93**		
4. COTS game	0.42	0.10	−0.02	0.03	−0.14	—	
5. Multiple-choice assessment	0.55	0.25	0.06	−0.25*	0.30*	0.02	—

**$p < 0.05$, **$p < 0.01$. M = mean, SD = standard deviation, COTS = commercial-off-the-shelf. Sex at birth coded as 0 = female, 1 = male. The COTS game scores and multiple-choice assessment scores are percentages. Estimates of reliability are bolded along the diagonal.*

Table 1.
Descriptive statistics and correlations for Studies 1–4.

4. Discussion

The purpose of this chapter was to provide information about the application of COTS games as game-based interventions. In the research series presented as part of this chapter, we sought to build upon a growing body of research literature on COTS games by showing the convergence between COTS game scores with traditional multiple-choice assessment scores on the same content. Overall, the research results answered our research question and showed that COTS game scores significantly and positively correlated with the traditional knowledge assessment scores in Studies 1–3 but not Study 4. Although a non-significant relationship was found in Study 4, we note that these results may be attributed to a limited sample size. Data collection was halted during Study 4 due to the onset of the COVD-19 pandemic and local laws prohibited in-person gatherings including the collection of in-person data which was part of this research protocol. Overall, these results suggest that when a COTS game is considered and chosen intentionally, that the COTS game scores may be reasonably similar to a traditional knowledge assessment score. We caveat these results by stating that these findings cannot generalize outside of these particular games and provide primarily a proof-of-concept to the consideration and use of COTS games. We encourage researchers and practitioners to collect and evaluate their own evidence for the COTS games they select and the intended application of those games.

4.1 Intentional game selection

The meaningful relationships between the COTS game scores and traditional assessments highlights the potential benefits of using game-based interventions. COTS games may provide viable solutions to learning and assessment contexts but must be considered carefully with the particular purpose of the game in mind. While some

COTS games may demonstrate meaningful uses without being intentionally designed to do so [7, 9], it is important to note that not all COTS games will demonstrate similar results. Here we highlight the importance of selecting the right COTS game for the intended purpose. To avoid potential pitfalls of using a COTS game, individuals can prioritize the selection process by incorporating time for planning and review into the early stages of their project. Having a thorough understanding of the differences in game design and content is critical to selecting an appropriate COTS game for the intended purpose. As mentioned earlier in this chapter, learning and assessment are examples of two desired outcomes. To further illustrate how a COTS game could be reasonably connected to the end goal, we provide more detail about the content of the COTS games used in the series of research studies presented in this chapter.

4.1.1 Study 1

In the game *Quintet*, the primary content focuses on players learning their distinct responsibilities (i.e. the Scientist role can survey the surroundings while the Engineer role can repair and boost various functions) while learning to communicate and coordinate their actions with their teammates. Because the game missions require teammates to coordinate the timing of their actions, gameplay inevitably promotes active communication and collaboration between teammates. Teams will automatically perform better in meeting the mission objectives when they are actively communicating and coordinating their different actions.

4.1.2 Study 2

In the game *Arm Surgery 2*, the player is presented with factual information about specific surgical equipment and procedures. During gameplay, players are guided through several procedures where they have to select the correct tool and perform the correct action before the game can progress. Progressing through the game naturally promotes active attention towards these tools and procedures as players must monitor and maintain the patient's health while performing their tasks to earn a higher score.

4.1.3 Study 3

In the game *PC Building Simulator*, players are responsible for their own PC repair shop. In order to improve their score, they must accurately diagnose the computer's problem, order the correct parts, and repair or replace those parts before returning the PC to the customer. When players aren't sure about particular parts, they can go to a guide to click on parts for the name and information on that part. For the gameplay to progress, players must continue to accurately identify and repair the defective components.

4.1.4 Study 4

In the game *Car Mechanic Simulator 2*, the player owns a car mechanic shop. Similar to the game in Study 3, the player must accurately diagnose the problem, order the needed parts, and replace those parts in the car. The individual can also go to schematics which will give the name and primary functioning of the different parts that will help them when trying to understand and diagnose a problem. Thus, gameplay naturally leads to improved understanding of the different parts of a car and their functioning.

Each of these topics are relevant to workplace learning and may present a relevant context for training or assessment. For example, in Study 1, participants must learn their role in the game but also communicate with members of their team, coordinate their actions, and effectively work together to meet the game objectives. In this instance, the participants cannot be successful together unless they work as a team. Although the environment of being in space does not present a typical work environment, players were actively engaging in workplace applicable teamwork skills. In contrast to this, the physical environment of the games in Studies 2–4 were relevant to the fields of their subject matter (healthcare, computer science, and automotive industry, respectively). As an increasing number of industries continue to consider and apply game-based interventions, the variety of different COTS game contexts becomes more relevant [41–43].

4.2 Low and high stakes contexts

While COTS games offer a reasonable alternative to the resource intensive commitment of developing a game, they do not provide the same precision and control compared to traditional methods, which can be an important component in some training and assessment contexts [9]. Because of this, it is important to only use COTS in particular contexts when this precision and control is not critical. Beyond assessing how the content of a COTS game is associated with the intended purpose, it's imperative to consider whether the data will be used in a low stakes or a high stakes scenario.

High stakes scenarios occur when results could have a large, consequential influence on a person's access to or receiving of resources. This means it might impact their pay, promotion, or selection into a program, course, or position. A low stakes scenario occurs when the results are used for general information given to the individual or for developmental purposes when shared with a larger institution. For example, using a game to assess applicants for a job would be high stakes. Whereas, using a game to assess high school students' language skills to provide feedback on how to better prepare for college would be a low stakes scenario.

An additional consideration to distinguish high and low stakes scenarios is when a cost is associated with the service and when a claim is made with regard to the actual result of the intervention. For example, someone paying for a training course intended to improve their skills with a software is high stakes and should have evidence that this program improves the skills in individuals that it claims to improve. Similarly, if the program is free but it claims to improve the skill, it should still provide this evidence based on its claim. In contrast, if the program is a free game intended for entertainment purposes and does not make claims about improving cognitive functioning or other skill development then it is low stakes.

High stakes scenarios should be approached with extra care and use instruments or interventions with strong psychometric evidence on the reliability and validity of intended use. To this end, we recommend COTS games to be used primarily in low stakes scenarios. Given the degree of importance and variation that can occur in high stakes scenarios, control over a game or gamified experience is paramount so that any bias or unintended negative consequences can be reconciled. COTS games are typically used without the option to customize aspects of the game, hindering the ability to identify and rectify any pitfalls. Using established and rigorous science-based processes for the development of game-based interventions is especially important for high stakes training or assessment purposes. GBL solutions are best developed following empirical instructional system design (ISD) methodology while GBAs should follow standard assessment development and validation processes to demonstrate the psychometric properties of the assessment [9].

4.3 Recommendations for software developers

Given the results of this paper, it appears that COTS games have potential uses beyond entertainment purposes when selected with intention and applied thoughtfully. Below are recommendations for software developers regarding actions which might benefit a COTS game for being used for alternative purposes.

Firstly, software developers can make a variety of game metrics available to collect from gameplay as providing frequent and detailed feedback on performance is a key component of both training and assessment [44]. When games are used for training or assessment purposes, having detailed metrics (e.g., how long a player spent on a level, how many incorrect selections were made etc.) can be a helpful device. COTS games which provide a metrics sheet or summary of data would greatly aid in the game's accessibility for alternate purposes. For example, many Massively Multiplayer Online Role-Playing Games (MMORPGs) and Multiplayer Online Battle Arena (MOBAs) make detailed game metrics publicly available for players to view statistics of their own and other players' performance to better understand and improve their future gameplay [45].

Secondly, when possible, it would help to have software developers create easily customizable options in their games that could increase the usability of the game for alternative purposes. For example, some games could add additional modes (e.g., single player, cooperative, or competitive modes) which allow for alignment with a greater variety of intervention purposes. Games could also provide more content related tutorials so players have the opportunity to learn about the context of the game in a tutorial rather than just learning through game mechanics. For example, a medical based game could provide specific information about tools, procedures, and different considerations when caring for a patient. Additionally, learning and assessment components can be an added benefit when thoughtfully integrated into other contexts such as educational settings [46].

Thirdly, when a game is naturally aligned to an area that might have potential for alternative uses outside of entertainment (e.g., game has content relevant to child or adult learning or the content might potentially assess a relevant topic area) it would help to have software developers seek input or feedback from potential users. This may include contacting educators to ask about particular features they would want in the game if they were to use it for learning or posting summaries of the game on online forums and asking for general ideas on how to improve the game if it were used for assessment purposes. Understanding, and marketing, the particular qualities and game characteristics that would be most beneficial for alternative uses would help in promoting the selection of the COTS game.

4.4 Understanding COTS games and modern technologies

Considering the rapid emergence of new technologies, it is relevant to consider how some of the most recent innovations in the gaming world might impact the results discussed here. New technologies such as XR and VR offer high fidelity experiences in contexts that can be created and customized for individuals to experience. The potentially high impact of these new experiences involves the degree of realism they create for the player. Several researchers have begun to explore these technologies and their potential uses for training and assessment purposes, highlighting the hyper realistic environment as a key feature that future researchers and practitioners can capitalize on [3, 27].

This realism can be a valuable training or assessment component of the game or experience for individuals. For example, if trying to teach medical students how to perform a particular procedure, it may be beneficial to have them perform this skill under a variety of conditions such as in a quiet, focused space but also in a loud, and busy space that might better replicate the environment of an Emergency Room. Using XR or VR, the procedure could be adapted to different scenarios, and embedded within a seemingly realistic ER environment that allows common sights and sounds from this experience to be included within the training session.

Similarly for assessment, it may be critical for an electrical engineer to be able to perform specific repairs on power lines and equipment. In XR or VR, these repairs could be assessed in an environment that emulates a heights scenario, where the participant truly feels they are at the top of a powerline. The sights, sounds, and sensations they experience can bring a sense of presence and embodiment that is difficult, if not impossible, to emulate in non-XR or VR environments. In short, it is clear that there are new technologies emerging that could further enhance the usability and application of future COTS games.

5. Conclusion

Research has demonstrated that some game-based interventions may provide benefits in learning and assessment contexts. As the research on game-based interventions continues to mature, more empirical evidence is being produced on how to design and implement games in a more impactful way [47, 48]. Game-based interventions have much potential [9], but game design can be resource-intensive, making COTS games an attractive alternative when the requisite time, resources, or expertise are not available for game development. In the focal research series, we sought to explore how COTS game scores converge with traditional knowledge assessment scores and found preliminary proof-of-concept evidence that in this context, three of the four COTS games demonstrated significant convergence with the respective traditional, multiple-choice assessment. Although these findings do not generalize outside the context of these particular games, they do provide a promising future to the further investigation of COTS games as viable solutions in low stakes instructional and evaluative contexts. Overall, we recommend COTS games within particular contexts where a well-suited game is chosen and encourage individuals to exercise intention in how the COTS games are used.

Conflict of interest

The authors declare no conflict of interest.

Author details

Diana R. Sanchez*, Amanda Rueda, Leila Jimeno Jimènez and Mahsa Norouzi Nargesi
San Francisco State University, San Francisco, USA

*Address all correspondence to: sanchezdianar@sfsu.edu

References

[1] Bedwell WL, Pavlas D, Heyne K, Lazzara EH, Salas E. Toward a taxonomy linking game attributes to learning: An empirical study. Simulation & Gaming. 2012;**43**:729-760. DOI: 10.1177/1046878112439444

[2] Landers RN. An introduction to game-based assessment: Frameworks for the measurement of knowledge, skills, abilities and other human characteristics using behaviors observed within videogames. International Journal of Gaming and Computer-Mediation Simulations. 2015;7:iv-viii. DOI: 10.4018/IJGCMS

[3] Sanchez DR, Weiner EJ, Van Zelderen A. Virtual reality assessments (VRAs): Exploring the reliability and validity of evaluations in VR. International Journal of Selection and Assessment. 2022;**30**(1):103-125. DOI: 10.1111/ijsa.12369

[4] Sanchez DR. Videogame-based learning: A comparison of direct and indirect effects across outcomes. Multimodal Technologies and Interaction. 2022;**5**:1

[5] Bai S, Hew KF, Huang B. Does gamification improve student learning outcome? Evidence from a meta-analysis and synthesis of qualitative data in educational contexts. Educational Research Review. 2020;**30**:100322. DOI: 10.1016/j.edurev.2020.100322

[6] Wu WH, Hsiao HC, Wu PL, Lin CH, Huang SH. Investigating the learning-theory foundations of game-based learning: A meta-analysis. Journal of Computer Assisted Learning. 2011;**28**:265-279. DOI: 10.1111/j.1365-2729.2011.00437.x

[7] Landers RN, Collmus AB. Gamifying a personality measure by converting it into a story: Convergence, incremental prediction, faking, and reactions. International Journal of Selection and Assessment. 2022;**30**(1):145-156. DOI: 10.1111/ijsa.12373

[8] Garris R, Ahlers R, Driskell JE. Games, motivation, and learning: A research and practice model. Simulation & Gaming. 2002;**33**:441-467. DOI: 10.1177/1046878102238607

[9] Landers R, Sanchez DR. Game-based, gamified, and gamefully designed assessments for employee selection: Definitions, distinctions, design, and validation. International Journal of Selection and Assessment. 2022;**30**(1): 1-13. DOI: 10.1111/ijsa.12376

[10] Ritzhaupt AD, Gunter E, Jones JG. Survey of commercial off-the-shelf video games: Benefits and barriers in informal educational settings. International Journal of Instructional Technology and Distance Learning. 2010;7:45-55. DOI: 10.1177/0047239515588161

[11] Stansbury JA, Wheeler EA, Buckingham JT. Can Wii engage college-level learners? Ease of commercial off-the-shelf gaming in an introductory statistics course. Computers in the Schools. 2014;**31**:103-115. DOI: 10.1080/07380569.2014.879791

[12] Sundqvist P. Commercial-off-the-shelf games in the digital wild and L2 learner vocabulary. Language Learning & Technology. 2019;**23**:87-113. DOI: 10125/44674

[13] Wiklund M, Ekenberg L. Going to school in World of Warcraft: Observations from a trial program using off-the-shelf computer games as learning tools in secondary education. Designs for Learning. 2009;**2**:36-55. DOI: 10.16993/dfl.18

[14] Sanchez DR, Nelson TQ, Kraiger K, Weiner EJ, Lu Y, Schnall J. Defining motivation in video game-based training: Exploring the differences between measures of motivation. International Journal of Training & Development. 2022;**26**(1):1-28. DOI: 10.1111/ijtd.12233

[15] Krath J, Schürmann L, Von Korflesch HF. Revealing the theoretical basis of gamification: A systematic review and analysis of theory in research on gamification, serious games and game-based learning. Computers in Human Behavior. 2021;**125**:106963. DOI: 10.1016/j.chb.2021.106963

[16] Armstrong MB, Landers RN, Collmus AB. Gamifying recruitment, selection, training, and performance management: Game-thinking in human resource management. In: Gangadharbatla H, Davis DZ, editors. Emerging Research and Trends in Gamification. Hershey, Pennsylvania, USA: IGI Global; 2016. pp. 140-165. DOI: 10.4018/978-1-4666-8651-9.ch007

[17] Sanchez DR, Langer M, Kaur R. Gamification in the classroom: Examining the impact of gamified quizzes on student learning. Computers & Education. 2020;**144**:103666. DOI: 10.1016/j.compedu.2019.103666

[18] Erhel S, Jamet E. Digital game-based learning: Impact of instructions and feedback on motivation and learning effectiveness. Computers & Education. 2013;**67**:156-167. DOI: 10.1016/j.compedu.2013.02.019

[19] Taub M, Sawyer R, Smith A, Rowe J, Azevedo R, Lester J. The agency effect: The impact of student agency on learning, emotions, and problem-solving behaviors in a game-based learning environment. Computers & Education. 2020;**147**:103781. DOI: 10.1016/j.compedu.2019.103781

[20] Yang YTC, Chang CH. Empowering students through digital game authorship: Enhancing concentration, critical thinking, and academic achievement. Computers & Education. 2013;**68**:334-344. DOI: 10.1016/j.compedu.2013.05.023

[21] Hummel HG, Joosten-ten Brinke D, Nadolski RJ, Baartman LK. Content validity of game-based assessment: A case study of a serious game for ICT managers in training. Technology, Pedagogy, and Education. 2017;**26**:225-240. DOI: 10.1080/1475939X.2016.1192060

[22] Shute VJ, Ventura M, Kim YJ. Assessment and learning of qualitative physics in Newton's Playground. Journal of Educational Research. 2013;**106**:423-430. DOI: 10.1080/00220671.2013.832970

[23] Stănescu DF, Ionită C, Ionită AM. Game-thinking in personnel recruitment and selection: Advantages and disadvantages. Postmodern Openings. 2020;**11**:267-276. DOI: 10.18662/po/11.2/174

[24] Kim YJ, Almond RG, Shute VJ. Applying evidence-centered design for the development of game-based assessments in physics playground. International Journal of Testing. 2016;**16**:142-163. DOI: 10.1080/15305058.2015.1108322

[25] Courtney L, Graham S. 'It's like having a test but in a fun way': Young learners' perceptions of a digital game-based assessment of early language learning. Language Teaching for Young Learners. 2019;**1**:161-186. DOI: 10.1075/ltyl.18009.cou

[26] Mavridis A, Tsiatsos T. Game-based assessment: Investigating the impact on test anxiety and exam performance. Journal of Computer Assisted Learning. 2017;**33**:137-150. DOI: 10.1111/jcal.12170

[27] Weiner EJ, Sanchez DR. Cognitive ability in virtual reality: Validity evidence for VR game-based assessments. International Journal of Selection and Assessment. 2020;**28**:215-236. DOI: 10.1111/ijsa.12295

[28] Auer EM, Mersy G, Marin S, Blaik J, Landers RN. Using machine learning to model trace behavioral data from a game-based assessment. International Journal of Selection and Assessment. 2022;**30**(1): 82-102. DOI: 10.1111/ijsa.12363

[29] Shute VJ, Wang L, Greiff S, Zhao W, Moore G. Measuring problem-solving skills via stealth assessment in an engaging video game. Computers in Human Behavior. 2016;**63**:106-117. DOI: 10.1016/j.chb.2016.05.047

[30] Westera W, Nadoski R, Hummel H. Serious gaming analytics: What students' log files tell us about gaming and learning. International Journal of Serious Games. 2014;**1**:35-50. DOI: 10.17083/ijsg.v1i2.9

[31] Levy R. Dynamic Bayesian network modeling of game-based diagnostic assessments. Multivariate Behavioral Research. 2014;**54**:771-794. DOI: 10.1080/00273171.2019.1590794

[32] Nikolaou I, Georgiou K, Kotsasarlidou V. Exploring the relationship of a gamified assessment with performance. Spanish Journal of Psychology. 2019;**22**:E6. DOI: 10.1017/sjp.2019.5

[33] Chan D, Schmitt N. Video-based versus paper-and-pencil method of assessment in situational judgment tests: Subgroup differences in test performance and face validity perceptions. Journal of Applied Psychology. 1997;**82**:143-159. DOI: 10.1037/0021-9010.82.1.143

[34] Outtz JL. The role of cognitive ability tests in employment selection. Human Performance. 2002;**15**:161-172. DOI: 10.1207/S15327043HUP1501&02_10

[35] Landers R. Developing a theory of gamified learning: Linking serious games and gamification of learning. Simulation & Gaming. 2014;**45**:752-768. DOI: 10.1177/1046878114563660

[36] Kim YJ, Shute VJ. The interplay of game elements with psychometric qualities, learning, and enjoyment in game-based assessment. Computers & Education. 2015;**87**:340-356. DOI: 10.1016/j.compedu.2015.07.009

[37] Sanchez DR. Videogame-based training: The impact and interaction of videogame characteristics on learning outcomes. Multimodal Technologies and Interaction. 2022;**5**. DOI: 10.3390/mti6030019

[38] Findley WG. A rationale for evaluation of item discrimination statistics. Educational and Psychological Measurement. 1956;**16**:175-180. DOI: 10.1177/001316445601600201

[39] Lord FM. The relation of the reliability of multiple-choice tests to the distribution of item difficulties. Psychometrika. 1952;**17**:181-194. DOI: 10.1007/BF02288781

[40] Sanchez DR, Langer M. Video Game Pursuit (VGPu) scale development: Designing and validating a scale with implications for game-based learning and assessment. Simulations & Gaming. 2020;**51**:55-86. DOI: 10.1177/1046878119882710

[41] Aziz R, Norman H, Nordin N, Wahid FN, Tahir NA. They like to play games? Student interest of serious game-based assessments for language literacy. Creative Education. 2019;**10**:3175-3185. DOI: 10.4236/ce.2019.1012241

[42] Fong G. Adapting COTS games for military experimentation. Simulation

& Gaming. 2006;**37**:452-465. DOI: 10.1177/1046878106291670

[43] Graafland M, Schraagen JM, Schijven MP. A systematic review of serious games for medical education and surgical skills training. British Journal of Surgery. 2012;**99**:1322-1330. DOI: 10.1002/bjs.8819

[44] Shute VJ, Ke F. Games, learning, and assessment. In: Ifenthaler D, Eseryel D, Ge X, editors. Assessment in Game-Based Learning. New York, NY, USA: Springer; 2012. pp. 43-58. DOI: 10.1007/978-1-4614-3546-4_4

[45] Steinkuehler C, Duncan S. Scientific habits of mind in virtual worlds. Journal of Science Education and Technology. 2008;**17**:530-543. DOI: 10.1007/s10956-008-9120-8

[46] Shute VJ, Ventura M, Bauer M, Zapata-Rivera D. Monitoring and fostering learning through games and embedded assessments. ETS Research Report Series. 2008;**2008**:i-32. DOI: 10.1002/j.2333-8504.2008.tb02155.x

[47] All A, Castellar EPN, Van Looy J. Assessing the effectiveness of digital game-based learning: Best practices. Computers & Education. 2016;**92**:90-103. DOI: 10.1016/j.compedu.2015.10.007

[48] Marquard JL, Zayas-Cabán T. Commercial off-the-shelf consumer health informatics interventions: Recommendations for their design, evaluation, and redesign. Journal of the American Medical Informatics Association. 2012;**19**:137-142. DOI: 10.1136/amiajnl-2011-000338

Chapter 3

Serious Games Development and Impact for Business Education

Dario Liberona, Aravind Kumaresan, Lionel Valenzuela and Giovanny Tarazona

Abstract

Learning methodologies and experiences have changed over the recent years thanks to the incorporation of digital technology, among many of these technologies are Serious Games, that has a better opportunity to be used during recent pandemic times, the process of designing and incorporating games technologies is not easy and there are very few available development tools, this paper focus on basic guidelines and a practical experience. Review the process of developing a serious game and address some of the challenges of making a serious game from scratch. The objective is also to understand the challenges of developing and implementing gaming mechanics in Serious Games and the impact and results of the experience of using it with students' samples from Latin America and Europe. The results of this study is that gamming has a very positive impact in the learning process of higher educations students, whom value the use of this technologies in their education, however in general these technologies are not being use in higher education, there is more simulation type implementations, especially in marketing and logistic areas. There is a need to train teachers and create pedagogical departments that will enhance and develop this experiential learning tools.

Keywords: serious games, edutainment, entrepreneurship, gaming base learning, startup education, gaming mechanics, startup journey game

1. Introduction

Higher Education is one of the key strategic factors that are helping to sustain and increase development and social mobility [1, 2], the world is rapidly changing, now we have almost complete generations that have play many games since very early childhood like generation alpha, meanwhile generation Z is mostly on University students and joining the labor force are super familiar with games, the first generation that had games and social networks since very early stage [3], generation Alpha is the first totally digital natives generation and already are playing games like never before, they will be joining universities around 2030 with more experience in digital gamming that any other generation (**Table 1**).

Computer games have already set their foundation on mainframe computers and then extended its use to personal computers and many devices and ways, electronic

IntechOpen

Generation Segment	Years Born	Workforce	Education Stage Now	Popular Games	Early Childhood Games (Under 11 years old mostly)	Characteristics	Formative Experiences	Aspiration	Learning Style
Generation X	1961–1980	35%	Mostly done with Higher Education	Pong (1972), Zork (1977), Space War, Space Invaders (1978) Galaxian (1979), Kung fu Master (1984), Pac Man (1980), Tetris (1984), Outrun (1986), Simcity (1989), Super Mario Bros (1988).	Pong (1972)		End of Cold War, Fall of Berlin Wall, Live Aid, Introduction of first PCs, Early Mobile Technology, Rising levels of Divorce	Work life balance	Participative
Generation Y/Millenials	1981–1995	29%	Graduate Level	Super Mario 64 (1997), The Legend of Zelda (1999), Pokemon Red/blue (1996), Sonic (1991), World Soccer (1994), Street Fighter (1992), Super Mario Kart (1992), Resident Evil (1996), Mind Craft	Pac Man (1980), Tetris (1984), Outrun (1986), Simcity (1989), Super Mario Bros 1988.		9/11 Terrorist Attacks, Play Station, Social Media, Invasion of Iraq, Reality TV, Geogle Earth,	Feedom and flexibility	Interactive
Generation Z	1996–2010	24%	High Schools, Undergraduate at Universitieis and Masters Degree	Technologies: Game Cube, Play Station 2, Xbox, Wii, Play Station 3, Xbox 360. Virtual Friends, Rez (2001), Wii Sports (2006), Call of Dutie (2007), Guitar Hero (2006), Silent Hill (2001), The Elders	Club Pinguin, Mini Juegos, Super Mario Bros, Tony Hack, The Sims, Wii sports, Halo, Tamagoshi (Digital Mascot), Legend of Zelda, Mario Kart, Where in the World is Carmen SanDiego,	Multitaskers, Not bound by work schedules, personal well being, lack of trust and skepticism. The largest generation Ever, fully global generation, connected trhough digital devices, and	Economic Downturn, Global Warming, Global focus, Mobile devices, Energy Crisis, Arab spring, Produce own media, Cloud computing, Wiki leaks	Security and stability, persue your passion	Multi-Modal

Generation Segment	Years Born	Workforce	Education Stage Now	Popular Games	Early Childhood Games (Under 11 years old mostly)	Characteristics	Formative Experiences	Aspiration	Learning Style
				Scroll (2006), The Legend of Zelda (2000), Halo (2001), The Sims (2000), GTA San Andreas (2004), League of Legends, Team Fortress 2, Counter Strike, Resident Evil, Grand Turismo, Need for Speed, Word of Warcraft, ForNite, Pokemon Go (2016)	Math Blaster, Barbie Detective, Rollet Coaster Tycoon, Sid Meirs Civilization,	egaged trhough social media. Digital integrators.			
Generation Alpha	2011–2025		Basic School		Technologies: Play Station 5, Xbox One, Nintendo Switch, Virtual Friends, Call of Dutie, Rock Smith, The Legend of Zelda, GTA 5, League of Legends, ForNite, Fall Guys, Getting Over it, Among us, Dauntless, Pokemon Go	Digital Natives, Entirely born and shape in the 21st century, and will probably be around the 22nd, more technologically literate generation ever.	Migration and diversity, Covid 19, ecommerce, Countries fail to Climate change, distance learning		Virtual

Table 1.
Characteristics of generations and most popular games (own elaboration).

games have been close to at least the so-called generations X, Y, Z many of generation Z are studding at universities now and soon generation Alpha born on this century will be joining higher education with a big exposure to gaming has never seen before. In the last 50 year we have seen tremendous growth in the gaming industry and since a decade or some gaming has become serious with a new branch of edutainment, new games with the purpose of helping to teach certain specifics concepts and decision-making experiences.

After generation Y some concepts of learning were incorporated into education, but it was generation Z the one that learned many concepts, math, sciences, languages, and others with interactive games mostly.

Table 1 shows the different generations and their most common games, we can clearly see how gaming has become more and more popular each decade, most of the generation Z and Alpha, the current and future higher education students have a culture of gamming and are familiar with diverse gaming technologies like never before, and generation Alpha has been using virtual reality and augmented reality, this generations have been very active in this 180 billion dollars industry.

Serious Games, also called educational games or applied game, Clark Abt is credited for coining the term "serious game" in the 1970s, defined as "games have an explicit and carefully thought-out educational purpose and are not intended to be played primarily for amusement" [4].

Digital games whose main purpose is "serious" in the sense that the learning outcome is more important than to simply entertain the players. The primary "serious" purposes can be to teach or train in specific areas providing learning experiences for students.

There are some video games that can be used for learning skills competencies, as they have an educational setting, such has Sims City, perhaps other areas such as Astronomy and Space-Themed Mobile Games.

However, these games need many video game design skills in order to be attractive for students, there are many categories of Digital games, simulations, virtual environments and mixed reality/media that have been applied to teach or train through responsive narrative/story, gameplay or encounters.

Nowadays constant innovation is needed in all industries, specially education, and the incorporation of serious games is a technology that has been evolving and helping to improve the students learning experience since the early 2000s, and has been slowly incorporated into university classrooms activities, there is still a big gap for teachers to adopt this new roles as game designers and game coordinators, they need to build new instructional design models that will benefit of both new technologies available and the new generations of students with long gamming experiences [5].

It is important to incorporate serious games and educational technology in order to teach skills and competences that are more practical and changing very rapidly, there is a need to give a more experiential dimension to the acquisition of knowledge from students, in this regards some games and digital simulations have been developed over the last decade [6–8].

The research group wanted to use this tool to improve the experiential learning of business development with a serious game denominated "Start up journey game", the objective of the game was to train and practice basic startup concepts.

The first steps were to determine what were the fundamental concepts to teach and then work with a specialized Game creation venture that will add the game experience best practices.

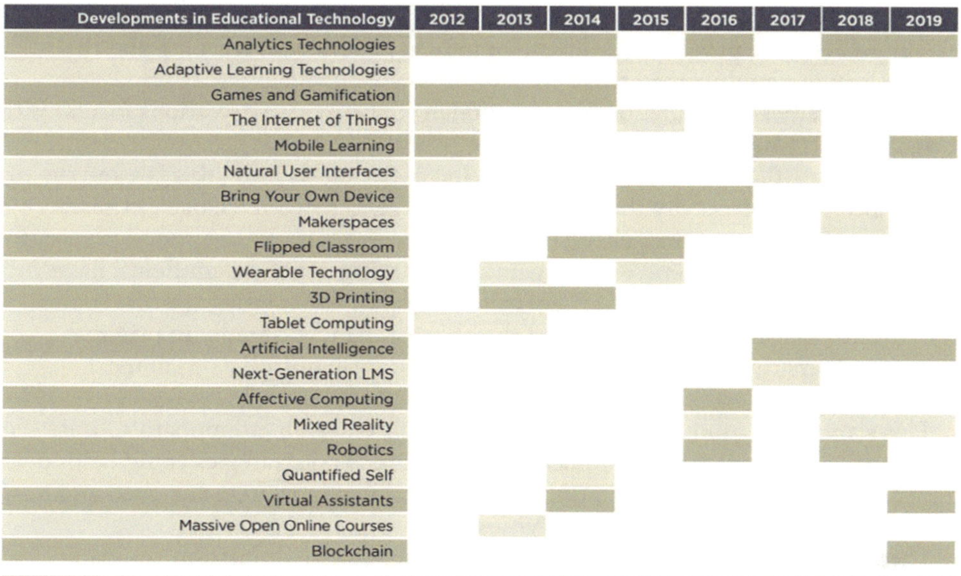

Figure 1.
Development in education technology trends. (report: 2019 higher education edition. Louisville, CO: EDUCAUSE).

In figure one we can review some of the educational technologies that have being incorporated in the last decade in higher education [1], here 21 new technological trends are identified, and we can see that games and gamification and many related technologies that complement this technologies such has analytics (that helps analyze students behavior and learning process), the use of mobile technologies and mixed reality. New technologies are being incorporated and the challenges of virtual education debt to the Covid pandemic has accelerated this incorporation (**Figure 1**) [9].

During this research we have made various surveys involving more that 375 students that have taking the Start Up Journey game in different countries and in different levels of higher education.

Some good practices applied to game developing and course implementation are review on this work.

It is not easy to produce full serious games, it takes time and could be very expensive, in the making there is two main challenges the content development in terms of the learning objectives and then what kind of gaming activities and type are the best to learn this, the gamification process that requires very specialized gaming concepts for users students that have a big expertise in gaming, but there are some companies developing simulations and tools that will help teaching specific contents.

2. About serious games

Games have been in society for a long time, there are evidence of dice playing since 3.000 ac, the go or Weiqi a board strategy game dates from 2.500 years ago played in China, many games had some learning concepts associated.

Serious and games seem concepts that are not normally related, but recently Serious games have proven to be effective to help experiential learning and that students can learn while they play.

"Serious games are games designed in which education (in its various forms) is the primary goal, rather than entertainment" [10–12].

Even though that there no agreed definition of Serious Games, this is often the case of new disciplines, basically the concept refers to games that have been primarily designed to learn rather than entertainment.

The field of edutainment, serious games or learning game design fields, have made big advances during the past years, they help understanding specific topics and acquire complex competences and they have expanded rapidly in primary schools, universities, and corporations, but they are still not very use in most Universities [8].

There are a few fundamental questions that this research attempts to address such as How has gamification been applied? What policies and strategies could help to successfully incorporate gamification, how is the learning process achieved and how students rate these serious games activities?

There are many Serious Games for higher education or simulation companies in the world today, such as:

Stratx Simulation (France), Cesim (Finland), EON Reality (USA), Simcoach Games (USA), Peak-Brainbow (United Kingdom), Costa Edutainment (Italy), and some new startups companies dedicated to develop games like Digital Dream Labs (USA), Game Lab (Chile), Fingerprint (USA),

We also have some traditional game base or simulation games such as Flight Simulator that was first launch in 1982, and now Microsoft Flight Simulation with a 2020 version, and is the longest running software from Microsoft, Pulse is a training serious game that simulates surgery on patients, Minecraft with an educational edition that was specially develop for schools.

Certainty COVID has been a catalyzer for technology adoption and distance learning in Higher Education since 2020 until 2022, but it has not really impacted Serious Games adoption jet, even that the need of engagement from students specially online and blended students is more necessary than ever.

3. Educational games (edutainment)

In **Table 2** we can see the most important concepts related to Serious games, even though they are related, their main purpose and learning objectives are different.

Educational games, can be traced back to the 1970's where "non digital" games where used in schools for math-related or social science understanding. The digital applications started around 1990's with the first multimedia computers, evolving the term to edutainment, but because of the poor quality and lack of understanding the interest decreased. Research has shown a positive effect of games as educational tools in various skills such as: strategic thinking, planning, communication, collaboration, group decision making, and negotiating skills [13, 14].

Edutainment, comes from "education" & "entertainment" it is designed to generate motivation, interest and a better understanding throw technology using games, music, internet or television to help both students and teachers in the process of learning. The market of business simulation for education are mainly centered on marketing, finance, strategy and optimization.

Category	Main purpose	Learning objective	Game characteristics	Application example
Gamification	Incorporate learning activities and modules.	Improve on a certain area.	Game elements	Prize for Results, Unlock activity levels
Serious Games	Build essential skills, games develop to learn specific objectives	Concepts, skills, competences, try experiential learning. Trainning value	All the game characteristics. More fun. Competition.	Flight Simulator
Edutainment	Entertainment	Basic learning concetps or skills	Few characteristics	Achive it with sesame street, Kahoot
Simulation Games	To simulate a real wrold scenario and desicion making impact	Learn impact of desicion on specific indicators.	Very Few game element. Sometimes competition and stages	MarkStrat
Training Games	Specific skills and competences	Team Work, communication skills, other specific.	Enternainment concepts	World Puzzle
Games	Entertainment	None	All the game Characteristics	World Puzz +C7:G13

Table 2.
Concepts related to serious games and edutainment. (own elaboration).

Gamification blends game mechanics with traditional eLearning activities and modules. For example using leaderboards, points, and eLearning badges add incentives for students to be more active on their online courses. Gamification and serious games are often grouped together they both motivate online learners and enhance their eLearning experiences [15].

4. Some facts about gamification

- Eighty percent of the learners say that their productivity would increase if the learning approach used in their university/organization is more game-like
- Eighty two percent of the learners are in favor of multiple difficulty levels and explorable content
- Seventy one percent of employees are not engaged for disengaged in their work
- Eighty nine percent of the learners show greater engagement of the LMS application has a point system
- Sixty two percent of the learners feel they would be motivated to learn if leaderboards were involved and they had the opportunity to compete with other colleagues

- Ninety percent of the learners recall information if the applying content within a stimulation

Source: Https://www.eidesign.net/gamification-in-elearning-facts/

5. Development of serious games

Serious games did not come into wide use until the 1990s with the PCs sales increased, even though many games were created before those years. At the time, educational games and other software evolved into "edutainment". However, interest in edutainment soon decreased, partly because of the (poor) quality of the games themselves and that playful experiences were not well accepted by higher education faculty that doubted the connection of entertainment and formal learning [16].

The problems encountered in edutainment are reflected in phrases such as "edutainment, an awkward combination of educational software lightly sprinkled with game like interfaces and cute dialog" [17], or "most existing edutainment products combine the entertainment value of a bad lecture with the educational value of a bad game" [13].

With the general renewed interest in serious games, game developers have moved from "skill and-drill interactive learning paradigms towards situational and constructionist approaches" [18]. Games in education is gaining acceptance, but their use is not widespread, and it is a controversial issue [18, 19].

Educational games is also faced with the challenge of providing research evidence of the acclaimed benefits, which currently is "complex and thinly spread", possibly because the study of games and gaming relates to several different disciplines; "as a result of the diversity and complexity of games themselves, and the range of perspectives taken by researchers, there are few hard and fast findings in the literature" ([14], p. 2; [20]).

Despite the "few hard and fast findings", research is showing positive effects of games as educational tools. Games can support development of a number of various skills: strategic thinking, planning, communication, collaboration, group decision making, and negotiating skills [13, 14]; see also Gee, unpublished manuscript). However, "hard facts and evidence" is for future research to provide. There is also a number of concerns to consider in order to realize the full potential of games as educational tools: resources (many schools have computers that are too old for new games, technical support, time for teachers to familiarize themselves with the game, etc.), how to identify the relevance of a game to statutory curricula, difficulty in persuading school stakeholders to the potential benefits of computer games, etc. [18].

5.1 Benefits of serious games

- Allow the decision-making process with little cost if failing or wrong, the possibility to fail and learn at a low cost. This also helps performance improving through try and error iteration.

- They are more experiential and interactive that lecture classes

- The freedom to experiment, games allow players to explore and discover new strategies and pieces of information
- The possibility to give quick feedback to students, and the possibility to analyze the student's decision making process.
- Problem base approach to learning, there are goals to achieve and problems to solve.
- The learning experience facilitates teamwork and collaboration among players.
- The freedom to assume different identities: games encourage players to see problems from a different perspective
- The freedom of effort: games allow players to go through periods of intense activity and relative inactivity, so that players can pause and reflect on tasks they have accomplished. To this end, gamification can be broken down into individual elements, each of which bring specific advantages and disadvantages to educational processes [21].
- Retain more information and learning stored in long term memory
- It is easy to monitor events and the advances of students, also keeps records of many situations.
- Possibility to establish emotional connection
- Multi tasking possibilities (**Figure 2**).

5.2 Serious games application challenges

The challenge facing serious games though, is to find a balance between the ludic and skills or knowledge transfer goals so that neither a dominant game mode

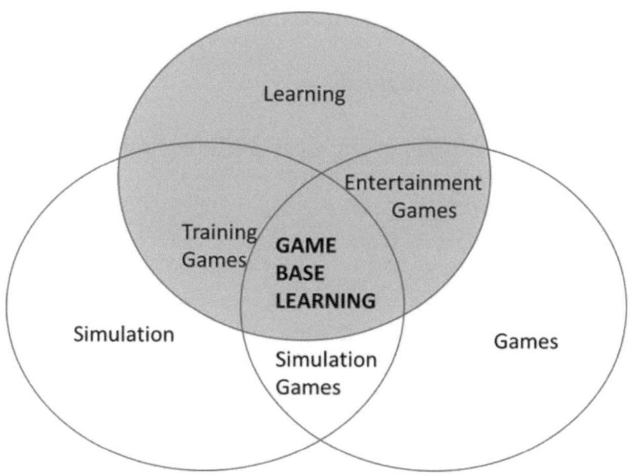

Figure 2.
Intersection of learning, computers science and games. Game-based learning according to Ulicsak and Williamson [22].

(taking away from the learning outcomes) nor learning mode (removing the fun element) is present.

6. Game development

You could develop your game, use and existing one or paid for a company that provides serious games alternatives like Gamelabeducation.com

a. Adapting an existing game: In the case of using an existing game, the cost tends to be lower, and the experience is good but is not personalize for the class.

b. Developing a game by the University/Professor: This alternative has various options, the two basic ones are to develop the game has part of project with the computer sciences department or engineering students that program it. Other option is to develop with a professional company does know about games and will develop a customize game for you. The cost of developing the game ranges between Us$ 60.000 to Us$ 300.000 according to the specific requirements and complexity, in any case will require many working hours from the teacher designing the game.

c. Use a serious game service that will provide the platform for you and you will use with a usual price per student that ranges between US$ 15 to US$ 125 dollars per student depending in the complexity of the service the prices have been soaring over the last few years, with an average close to 50 dollars per student in licenses.

Some steps to work on the project are:

1. Goal Setting is the first step (What do you want to do and learning Objectives);

2. Pick The Right academic and developer partner team;

3. Focus On eLearning Character Development;

4. Identify The Ideal Game Mechanics;

5. Embed Third-Party Resources to Increase Serious Game Interactivity;

6. Design a Prototype;

7. Test it with students;

8. Develop a logic teaching experience;

9. Monitor de learning process.

The time use to develop a game will range between 6 months (if you are customizing) and more than two years, an average (according to the survey) is between 12 and 18 months.

The game usually is adapted to different academic level of students, most them can be use in High School, Undergraduate, Senior Undergraduate and master's degree programs.

7. The startup journey entrepreneurship game

7.1 Goal

Teaching Entrepreneurship and Creation of New Businesses were the main goal of the Serious Games.

After the evaluation of the contents, various approaches, we came to the conclusion that it was extremely difficult to develop a game that could help you train the expertise and experiences, decision making of an Startup Journey, usually lasting 5 to 6 years in most countries in the world, special cases like silicon valley you could speed up the process but it is difficult.

The academic training team agreed in that the experiential teaching was possible in 3 different games, or game levels, one that will happen the first year of setting up the enterprise, the resulted game will be happening on the month 3rd to 12 of the first year of launching the company and the main learning objectives are:

7.2 Market fit

Through market research the players will find information about the customers segments in the objective place, and will define the segment and two main characteristics of the developed application.

7.3 Competitors analysis

Through market research, developing networks and visiting places the players will gather information on the Competitors, and what are they offers in the market plus size.

7.4 Network development

The players could develop their networks through the hired personal, sharing with them through coffees and telephone calls.

7.5 Team development

A right team is important for any startup so it is important to recruit and select the right team for the project.

Through the simulation, players will understand the importance of the following aspects when starting a business: formation and strengthening of a work team, creation of a network of contacts, benefits of having a mentor, identification of the needs of each target segment and positioning of the product with respect to the competition and having a basic analysis of the projected profitability through cash flow information.

8. Pick the right academic and developer partner team

The academic team for the game were mostly experts in startups, it consisted on 3 main consultants, one with 26 startups experience and three exits, the second with 9 startups experience and academic experience with a Phd, and the third with a long academic experience in entrepreneurship teaching and research.

Regarding the software development team, the agreement was with gamelab education and startup company that had developed two other logistics games and had experience with higher university teachers.

Also the team counted with English and portugues speakers.

8.1 Focus on eLearning character development

The game was though to be applied in presence and e-learning mode and was decided to be develop in at least three main languages, English, Spanish and Portuguese (**Table 3**).

8.2 Identify the ideal game mechanics

There are common types of game mechanics and experiences such as.

The logic chosen for the game was MMO – RPG, Masive multiplayer online with role playing, so the whole class could be part of the game, and you could have a small class of 20 players in groups of 3, or a large class of 100 players with groups of 4 students each.

The players could play in teams or alone, but the academic recommendation is that there are teams of 3 or 4 participating and playing in parallel. The basic role is of the startups founder team, they will prospect the market and take decisions.

There are places that are open after achieving certain goals, and an startup mentor that is achieved when the players develops the network (**Figure 3**).

Mechanics may also consist of incentives or rewards, such as badges that unlock new content or leaderboards that fuel their learning motivation.

Type of game	Description	Examples
FPS	First person Shooter	Doom, Call of duty
RPG	Role playing games	The Witcher, Fable 2, Mario Kart 8
MMO	Masive Multiplayer Online	Audition, Fornite
MMO RPG	Masive Multiplayer O. Role Playing	World of Warcraft, Fantasy line, Sim City
MOBA	Multiplayer Online Battle Arena	League of Legends
RTS	Real Time Strategy	Warcraft III, Rise of Nations
ADV	Adventure Game	The legend of Zelda, Fall out, Asassin Creed
FPS/RPG	Hybrid Game	Halo infinite

Table 3.
Types of video games and examples (own elaboration).

Figure 3.
Startups game scenery ecosystem (own elaboration).

9. Embed third-party resources to increase serious game interactivity

What do you do if you want to increase online learner engagement and immersion, but you do not have the time or resources to develop additional content?. You can enhance the game by embedding third-party resources, you can have add ons for your serious game without having to create everything in-house. Include a link to a helpful tutorial that walks online learners through a task, or embed a YouTube video that explores a topic at length. You can even create printer-friendly eLearning courses by integrating PDF files and other downloadable documents that online learners can access via the serious game.

10. Design a prototype

Choosing your eLearning authoring tools carefully, integrating the most effective game mechanics, and knowing when it's time to outsource are the secrets to serious game success.

A light version of the game was developed with most of the functionality and then got the feedback for testing players helped increase and solve many of the serious games improvement opportunities (**Figures 4** and 5).

11. Test it with students

This is an important step in order to learn how to teach and implement the gaming experience.

You learn aspect such has, you need to do short trial test of the simulation first and after training rounds, in this case two training rounds are suggested before the final round.

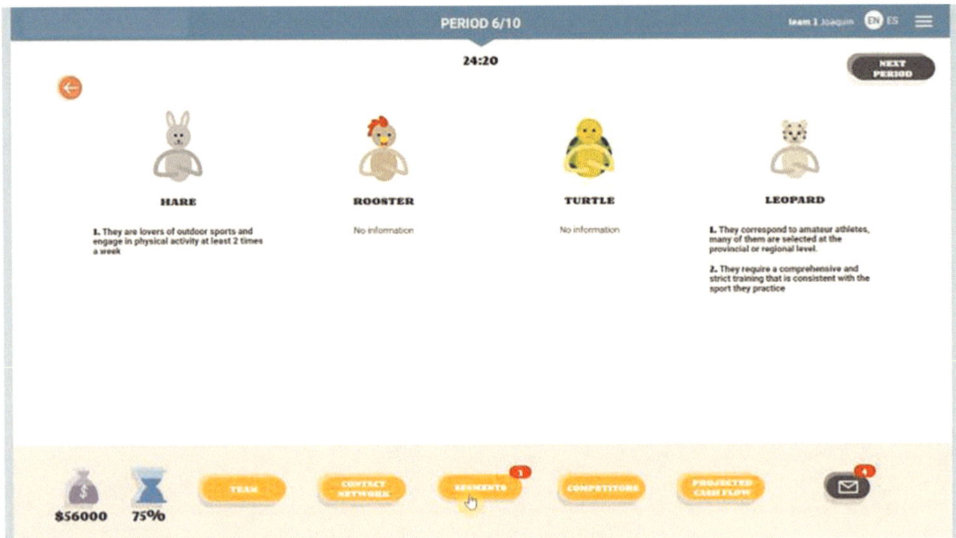

Figure 4.
Segmentation characters and basic characteristics (own elaboration).

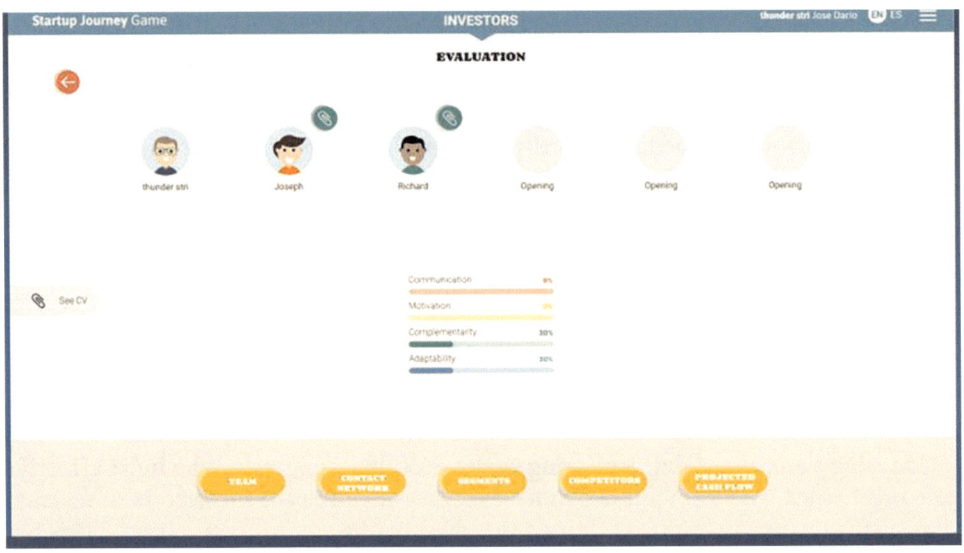

Figure 5.
Team selection and creation example, with skills (own elaboration).

12. Develop a logic teaching experience

Once the game has been developed and tested it is important to create the learning experience. The game mechanics could be taught and teach in many different ways so the teaching experience could be organize in many ways, depending on the audience and on the program that is being tough, usually it has to be personalize to the audience, undergraduate versus postgraduate, short programs, long programs.

Prizing rewards for the winners is a clear motivation for the players.

13. Monitor de learning process

Surveys could help to learn about the students' learnings, likes, and experience.

The gaming should have a logs system that tracks the players decisions, times involved, sequence of plays and other relevant information.

14. Results on classrooms

The game was use in four different classes, two hundred and fourteen college students and one hundred and seventy-five senior high school students on pre-college workshops were interviewed and participates with the following results (**Tables 4–7**).

If educating the player should be the primary goal of serious games like Michael and Chen [19] proposed, the results on the learning process was clearly achieved, the students got better results than just preparing their classes, the students were able to share their experiences and discuss the results, plan strategies and. Class attendance also increased in classes that involved game simulations close to 95% (**Figures 6** and 7).

In **Figure** 7 we can observe the story telling of the game, firs the students learn how to go through the city spaces and take the decisions they need in order to learn about the customer segments (Market research), then build a team

Have you participated in serious games before (university)		Have you participated in serious games before (high school)	
No	57%	No	77%
Something similar	17%	Something similar	11%
Yes	27%	Yes	27%

Table 4.
Have you use serious games before (own elaboration).

Which learning method do you prefer		Which learning method do you prefer	
reading papers and text books	**13%**	**Reading papers and text books**	**11%**
Presentation Classes &Videos (Regular Teacher Class)	18%	Presentation Classes &Videos (Regular Teacher Class)	16%
Online courses – E-learning	10%	Online courses – E-learning	4%
Serious Games - Simmulators	17%	Serious Games - Simmulators	27%
Workshops	12%	Workshops	16%
Study Alone	14%	Study Alone	9%
Personal/Private Teacher	2%	Personal/Private Teacher	2%
Reading Books	2%	Reading Books	2%
Videos - Youtube	14%	Videos - Youtube	13%
Other (please specify)	0%	Other (please specify)	0%

Table 5.
What is your preferred method of learning (own elaboration).

What is the contribution of the Serious game to learning (University)		**What is the contribution of the Serious game to learning (High School)**	
Very Low	**0.0%**	**Very Low**	**0.0%**
Little	2.8%	Little	3.1%
Medium	26.8%	Medium	19.8%
High	32.4%	High	29.2%
Very High	38.0%	Very High	47.9%

Table 6.
Evaluation of the serious game experience (own elaboration).

Did the game contributed to learning about Entrepreneurship		**Did the game contibuted to learning about Entrepreneurship (High School)**	
No contribution at all	**0%**	**No contribution at all**	**1%**
Learned some new concepts	33%	Learned some new concepts	8%
It made learning easier and fun	24%	It made learning easier and fun	34%
A much better understanding of the concepts	29%	A much better understanding of the concepts	35%
Really helped to clarify and team entrepreneurship concepts	14%	Really helped to clarify and learn entrepreneurship concepts	22%

Table 7.
How was the level of contribution to learning with the serious games (own elaboration).

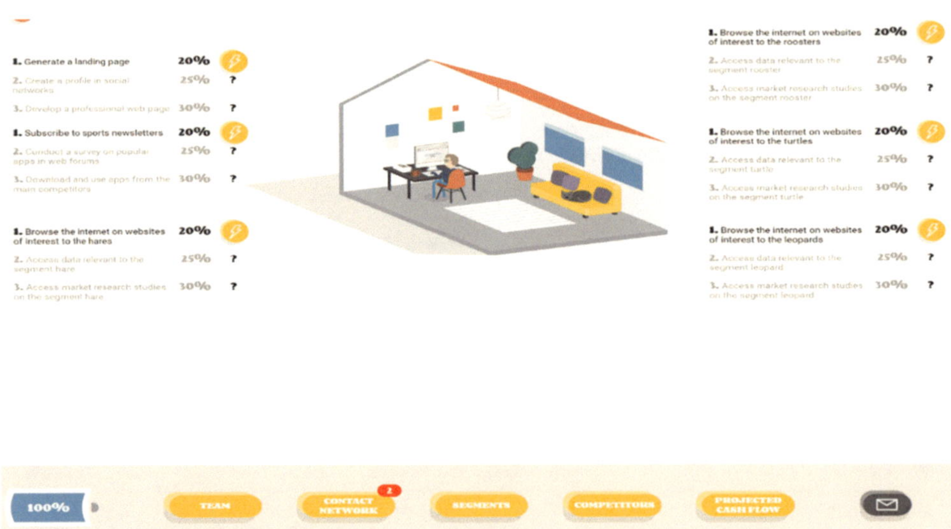

Figure 6.
Startup journey game (Universidad Tecnica Federico Santa Maria/Gamelab, Chile.

(Recruiting, interacting, motivating, complementing skills), through events and participating in activities they will learn about their competitors (Competitors map), then they will be able to find some financial information about the segments (Profitability and market potential), with all the segments information they will go

Figure 7.
Startup journey serious game logic (own elaboration).

to the Final decision making stage, they will decide what market segment to address and with what kind of application and functionality focus (Market Fit), with this the game ends, the investors with invest certain amount in the Startup or not invest and will give feedback to the Startup team. The group that raises mor of the capital wins.

15. Conclusions

Certainly, there has been a change in the culture and the way young students learn, all millennials are used to play digital games since very early age, some of them are role playing games such as Oblivion, the Whitcher, Pillars of Eternity, South Park, Fornite, so they are more ready for Serious Games than previous generations.

The fast paste of technological changes, business model changes, society changes make more challenging the training and education of young professionals, serious games are an excellent tool to make the experiential learning easier, to incorporate new technologies to the learning process, and to shorten the learning process, the experience of flight simulator has a training tool has slowly open the path to the use of innovative learning technologies, however adoption has been slow so far.

The process of creating a learning game is indeed very complex but there are many alternatives in the market for using already develop games.

The Universities will benefit of developing areas that will help them produce serious games in a serial format, with specialist teams or partnering with game developing companies that have already developed simulations or have the ability to personalize games.

New technologies will be incorporated to serious games and gamification, virtual reality, face and voice recognition, gesture control, new graphics, augmented reality, wearable gaming, on demand gaming are some examples of the new technologies that will help improve the learning experience and will deliver more engagement and knowledge acquisition [23].

Students really value experiential learning through serious games, even though they are demanding more quality and complexity, usually have the benchmark of games that have had years on the making and multiple hundred of dollars in development, anyways with basic gaming the students are engaged and declare that they do learn more concepts and enjoy the classes more.

The cheaper availability new technologies such as Computer Graphics, virtual reality and interactive visual simulation, hardware interactive sensors and tools, human computer interaction, will provide new grounds for Serious games development that could be highly beneficial both for higher education and industry.

During 2020–2021 since the pandemic outbreak the need of developed virtual entertaining learning programs has increased, the video game industry is close to 150 Us$ Billion and growing, more than 2300 developing companies, 70,000 employees are part of the new generation of games, in the Universities and higher education there is a long way to go forward but there will be rapid growth and adoption in the next decades and the incorporation of serious games will be also part of the most early education. Probably the Higher Education Industry with these new technologies will have more changes in the next decade that it has in the last 2000 years.

Author details

Dario Liberona[1*], Aravind Kumaresan[2], Lionel Valenzuela[3] and Giovanny Tarazona[4]

1 Department of Business and Culture, Seinajoki University of Applied Sciences, Seinajoki, Finland

2 Department of Industrial Management, Vaasa University, Vaasa, Finland

3 Departamento de Ingeniería Comercial, Universidad Tecnica Federico Santa Maria, Valparaíso, Chile

4 Universidad Distrital Francisco Jose de Caldaz, Bogotá, Colombia

*Address all correspondence to: dario.liberona@gmail.com

References

[1] Alexander B, Ashford-Rowe K, Barajas-Murphy N. EDUCAUSE Horizon Report 2019 Higher Education Edition. Louisville, CO: EDUCAUSE; 2019

[2] Bernal LLR, Medina VH, López Quintero JF. Knowledge management and intellectual Capital in a University Context. Lecture Notes in Business Information Processing. 2015;**2015**(224): 741-753

[3] Tabaka M, Gen Z. (2021). Will Make Up 24 Percent of the Global Workforce in 2020. Here's What Employers Need to Know. www.inc.com Newsletter. Available from: https://www.inc.com/marla-tabaka/gen-z-will-make-up-24-percent-of-global-workforce-in-2020heres-what-employers-need-to-know.html [Accessed: December 18, 2021].

[4] Edutechwiki (2021). Serious game. Available from: http://edutechwiki.unige.ch/en/Serious_game. [Accessed: November 28, 2021]

[5] Christian Y, Christian L. Integrating Serious Games in Higher Education Programs. Ramallah, Palestine: Paper presented at Academic Colloquium 2010: Building Partnership in Teaching Excellence; 2010

[6] Beranic T, Marjan H. Introducing ERP concepts to IT students using and experiential learning approach with and emphasis on reflection. Sustainability. 2019;**11**:4992. DOI: 10.3390/su11184992

[7] Giannakou I, Paraskeva F, Alexiou A, Bouta H. Social and emotional development for teachers in second life: Emotions & Connections are ubiquitous. In: Communications in Computer and Information Science Learning Technology for Education in Cloud–The Changing Face of Education. Vol. 620. Springer Series; 2016. DOI: 10.1007/978-3-319-42147-6_4

[8] Liberona D, Rojas C. Entrepreneurship knowledge transfer through a serious games platform. In: Communications in Computer and Information Science. Vol. 731. Springer International Publishing; 2017. pp. 110-122

[9] Khlaif ZN, Salha S, Kouraichi B. Emergency remote learning during COVID-19 crisis: Students' engagement. Education and Information Technologies. 2021:402-405. DOI: 10.1007/s10639-021-105664

[10] Oxford Analytica. Gamification and the Future of Education. 2016. Available from: https://www.worldgovernmentsummit.org/api/publications/document?id=2b0d6ac4-e97c-6578-b2f8-ff0000a7ddb6 [Accessed: December 10, 2019].

[11] PC Gamer. 2019. The Best RPGs on PC. Available from: https://www.pcgamer.com/best-rpgs-of-all-time/. [Accessed: January 4, 2020].

[12] Susi T, Johannesson M, Backlund P. 2007. Serious Games–An Overview. Available from: http://scandinaviangamedevelopers.com/downloads/HS-IKI-TR-07-001_PER.pdf

[13] Squire K, Jenkins H. Harnessing the power of games in education. Insight. 2003;**3**(1):5-33

[14] Kirriemuir JK, McFarlane A. Use of computer and video games in the classroom. In: Proceedings of the Level up Digital Games Research Conference. Netherlands: Universiteit Utrecht; 2003. Available from: http://www.silversprite.com/

[15] Freitas S De, Liarokapis F. (2011). Serious Games: A New Paradigm for Education ? Introduction: Serious Games: A New Paradigm, 9–23. DOI: 10.1007/978-1-4471-2161-9

[16] Thomson BOA, Yakimenko E. Edutainment as a modern technology of education. Procedia-Social and Behavioral Sciences. 2015;**166**(2015): 475-479

[17] Zyda M. From visual simulation to virtual reality to games. Computer. 2005; **38**(9):25-32

[18] ELSPA. Unlimited Learning: Computer and Videogames in the Learning Landscape [White Paper]. Royaume-Uni: ELSPA (Entertainment and Leisure Software Publishers Association); 2006. Available from: https://www.org.id.tue.nl/IFIP-TC14/documents/ELSPA-report-2006.pdf [Retrieved: 16 November 2016]

[19] Michael DR, Chen SL 2006. Serious games: Games that educate, train, and inform

[20] Brian Solis (2014). The Future of Learning Is Stuck in the Past: Why Education Is Less about Technology and More about Behavior. Available from: http://www.briansolis.com/2014/03/future-of-learning-takes-learning/ [Accessed: December 10, 2016].

[21] Catalano C, Angelo L, Mortara M. Best practices for an effective design and evaluation of serious games. Semantic Scholar. 2014;**2014**

[22] Ulicsak M, Williamson B. Computer Games and Learning. Bristol Future Labs. A Future Lab Handbook. 2010. Available from: https://www.nfer.ac.uk/publications/futl01/futl01.pdf [Accessed: 16 December 2021]

[23] Xu F (2013). The formation and development of Ikujiro Nonaka's knowledge creation theory. In: G. von Krogh et al. (Eds.), Towards Organizational Knowledge: The Pioneering Work of Ikujiro Nonaka (pp. 60-76). Basingstoke, UK: Palgrave Macmillan.

Chapter 4

The Use of Serious Games for Learning Cardiopulmonary Resuscitation Procedures: A Systematic Mapping of the Literature

Ingrid Nery Mendes, Maicon de Araújo Nogueira, Filipe Valente Mendes, Otavio Noura Teixeira and Viviane Almeida dos Santos

Abstract

This article presents a systematic mapping, with an analysis of 35 selected works according to established criteria, seeking to connect the points and find relevant information for the following research areas: basic life support, cardiopulmonary resuscitation, serious games, and games for healthcare. Among the main results found, we can mention the representativeness of works by regions and their most productive years, the most common platforms, noting a focus on VR technologies, in addition to identifying the preference for the Unity 3D tool for implementations. It was also possible to show that serious games can be very effective in teaching CPR.

Keywords: cardiopulmonary, resuscitation, CPR, basic life support, BLS, virtual training, virtual reality, VR devices, serious games, VR, systematic mapping, healthcare

1. Introduction

The basis of an electronic game is nothing more than a set of rules to be assimilated and through the repetition of their actions, the player can master the proposed challenges printed in the game mechanics and so advance through the stages, improving your skills, with the objective of winning the final challenge. According to Coelho and Nunes [1], in education, games have two functions: a playful and an educational one.

Playfulness refers to moments of leisure and fun, while education is portrayed through the acquisition of knowledge of the game content and understanding of the player's world. Games also provide the development of social skills, attention, concentration, logical reasoning, and others. The Serious Games (or Serious Games) bring the proposal to evolve and expand the concept of Educational Games. With a focus on training and management applied through game mechanics, they aim to deal with issues relevant to society, with the aim of generating effects outside the virtual domains, adding value to the player's individual knowledge in the real world.

The Serious Games (or Serious Games) bring the proposal to evolve and expand the concept of Educational Games. With a focus on training and management applied through game mechanics, they aim to deal with issues relevant to society, with the objective of generating effects outside the virtual domains, adding value to the player's individual knowledge in the real world.

Due to this, initiatives such as Games for Health [2, 3] and Games for Change [4] emerged, linked to this end and which also contribute to a promising field in the field of research, Game Studies [5]. With the popularization of digital games and the aging of the generations that grew up playing this product from the entertainment industry, this market trend tends to increase more and more, according to Vasconcellos et al. [6]. It is noteworthy that the theme of Games and Health is relatively new in academic events, such as SBGames—Brazilian Symposium on Games and Digital Entertainment, which at the end of 2022 will reach its 21st edition. Since 2016, a Workshop has been held internally in the area of Health, which last year became an official track for the dissemination of research and the development of games in this area [7]. At an international level, we have the International Conference on Serious Games and Applications for Health—SEGAH, which has been in operation since 2012, is in its 10th edition, and is an event affiliated with the IEEE [8].

According to the document from the Brazilian Archives of Cardiology, by Gonzalez et al. [9], training in Basic Life Support (BLS) generates individual specific skills that can be forgotten in a short time. Due to the lack of opportunities to make use of the procedures, or even the absence of constant practices, this knowledge has an average expiration time of 3–6 months. Therefore, there is a real need to generate maintenance of this knowledge, through simplified training for laypeople, in order to increase the durability of the retention of this knowledge in people's minds. The games can be used at any time, without an appointment, date, or place, just having a computer, a game console, or a mobile device and the willingness to practice. By applying game mechanics that have rules and that, in order to be mastered, need to be repeated in a dedicated way, the use of digital games helps in memorization, fixation, or recall of dormant knowledge. Thus, the union of learning BLS procedures with electronic games is a possibility to help spread this important knowledge, capable of saving many lives.

The motivation for this study comes from the interest of the author in games, together with the possibility of contributing to the research of the doctoral student in Nursing Maicon Nogueira [10], as there is a mutual interest in the union of these areas of research. Thus, this work aimed to analyze the production of academic publications related to health, focused on CPR in the last 5 years from the bases of SBGames, IEEE, Portal da Capes, and Google Academic, in order to identify and analyze the characteristics presented, to get an overview of what has been done in the area.

After this introduction, this chapter is structured where Section 2 brings the theoretical foundation of our objective and its relevance. Section 3 addresses the methodology used to compose this study with the selected articles chosen. Section 4 shows a variety of results that could help us to understand the extension of this theme and how others are approaching it, finally, Section 5 concludes this analysis followed by all the references used in this work.

DOI: http://dx.doi.org/10.5772/intechopen.102399

2. Theoretical foundation

2.1 Serious games

According to Derryberry (2008), what separates serious games from the rest of games is the focus on a specific and intentional learning result to achieve serious, measurable, and continuous performance and behavior changes [11, 12]. The repertoire of computing strategies for medical education is becoming more comprehensive, with the introduction of virtual learning applications (e-learning), based on games, gamification, and via mobile devices. A variety of serious games is being used more often in health education, bearing in mind that health care students are young and technology-savvy. The increased interest in games is evidenced by the growing number of reports and systematic reviews on the use of games in education. According to Bergeron [13], Serious Games is understood as an interactive computer application, with or without a certain connected hardware component, which has a challenging objective, is fun to play, has some scoring concepts, and adds a skill to the user, knowledge or attitude that can be applied in the real world. Games are called serious when they have a pedagogical purpose [14].

2.2 BLS and cardiopulmonary resuscitation

Basic Life Support (BLS) consists of a set of steps and maneuvers performed sequentially, which include immediate assessment and intervention in each phase of Cardiopulmonary Resuscitation (CPR), identified as follows: C—circulation (evaluation of signs of circulation and performance chest compressions), A—airway opening (assessment and correct positioning of the airways), B—breathing (assessment of respiratory movements and performance of ventilation), and D—early defibrillation. These recommendations are based on the guidelines of the International Liaison Committee on Resuscitation (ILCOR) and the international scientific consensus of the American Heart Association (AHA) [10].

3. Methodology

Carrying out systematic mapping of the literature, according to Kitchenham and Charters [15], requires the creation and completion of a protocol, which consists of detailed planning, which comes with the collection of all the necessary data to conduct the construction of the review. The steps to be performed are these:

- Define research questions
- Define search keywords
- Define the sources (bases) to be consulted
- Define the types of works that will be part of the research
- Define work languages
- Define the quality criteria for primary studies

- Define the search string
- Define inclusion and exclusion criteria
- List included and excluded articles
- Start the process of filling out the data extraction form, which will be carried out for each of the selected articles
- Conduct analysis of results: grouping, comparing, and critically discussing related works
- Define the existing indicators, metrics, criteria, and gaps

In addition to creating the protocol file, a tool to support the management of literature reviews was used. The chosen one was the national START (State of the Art through Systematic Review), authored by UFSCar/LAPES, in version 3.0.3 beta [16].

3.1 Survey questions

Following Kitchenham and Charters's study [15], and in order to find material related to the proposed topic, the following research questions were created:

- Q1: How have serious games been used to teach CPR protocols?
- Q2: How effective are serious games in teaching CPR?
- Q3: What is the state of the art of serious health-focused games for teaching CPR?
- Q4: What platforms are used to experience the games?

3.2 Strategies used in search engines

To continue the research procedure and obtain the answers to the questions, the keywords listed below were defined:

Keyword	Synonym
Serious games	
CPR	
Resuscitation cardiopulmonary	Cardiopulmonary resuscitation
BLS	

Based on these keywords, the following search string was defined at first: "serious games" OR "serious games" AND "RCP" OR "cardiopulmonary resuscitation" As the search return appeared to be too broad (187 items in the Google Scholar and 221 in Capes Journals), in addition to presenting several articles not related to the research interest, the string was redone.

The new string took the following form: ("serious games" OR "serious games") AND ("RCP" OR "cardiopulmonary resuscitation" OR "Cardiopulmonary Resuscitation" OR "CPR") AND ("SBV" OR "basic life support" OR "BLS" OR "Basic life support"), returning 110 items in Google Scholar.

On the Capes Journal Portal, the string had to be adapted, as it presented incompatibility with the previous format, obtaining the following form: (serious games) OR (serious games) AND CPR OR BLS OR ((Cardiopulmonary Resuscitation) AND (basic life support)), returning 176 items. After some observations in the listed articles, it was noticed that certain terms should be excluded from the string, as they have the same acronym of BSL and CPR, they were integrated in the search result.

Then, after a new adjustment to the string, this format was agreed to: (serious games) OR (serious games) AND CPR OR BLS OR ((Cardiopulmonary Resuscitation) AND (NOT (Common-Pool Resource)) AND (NOT (biomedical laboratory) science))), with the option to return only peer-reviewed journals, which resulted in 76 items.

The databases that returned in the search performed within the Capes Portal were PubMed, Scopus, Advanced Technologies & Aerospace Database, Directory of Open Access Journals—DOAJ, Web of Science, SpringerLink, Sage Journals, ERIC, and ACM Digital Library.

A private search was also performed in the IEEE database, which returned 21 items with the adapted string: ("All Metadata": "serious games") AND ("All Metadata": "Cardiopulmonary Resuscitation") OR ("All Metadata": "CPR training") OR ("All Metadata": "basic life support").

Finally, in the SBGames repository, which are the Annals of the Symposium, the search was performed manually, having found four items that included the chosen keywords. And this result was manually entered into START. BIBTEX files were also generated with the results of each search performed in the databases (Portal da Capes, Google Academic, and IEEE) to be inserted in an automated way in the support tool for the management of systematic literature reviews.

Searches were carried out between February and July 2021, through the academic registration valid on the CAPES portal and the mapping database, extracted during this period, and the final selection of chosen studies was systematized through MS Excel spreadsheet, which later also contributes to the elaboration of the graphs for this work. Also, during this period, learning the technique of Systematic Mapping was contemplated, through the subject of Scientific Methodology, which was administered by Professor Viviane Santos, collaborator of this work.

3.3 Inclusion and exclusion criteria

This item was also developed based on Kitchenham and Charters [15] with the formulation of the following definitions below, presented in **Table 1**:

In order to exclude erroneous terms that would invalidate the search results within the consulted databases, this list of criteria was created. In particular, the terms Common-Pool Resource/Cascaded Pose Regression (CPR) and Biomedical Laboratory Science (BLS) were included because they have the same abbreviation as the terms Cardiopulmonary Resuscitation (CPR) and Basic Life Support (BLS), respectively.

3.4 Criteria for selection of studies and identification of qualities

To obtain the works that contained the object of study of interest to the researchers in this review, the criteria were as follows: (a) whether the study uses volunteers to test

Criteria	Results
Being within the time frame of the last 5 years (2016–2021)	IC-01
Belonging to the disciplines of nursing, computer science, and medicine (specifically on the Capes Portal)	IC-02
Contain relationship with cardiopulmonary resuscitation (CPR)	IC-03
Academic works in English and Brazilian Portuguese	IC-04
Be from the physical sports spectrum (games)	EC-01
Contain the term Common-Pool Resource or Cascaded Pose Regression (CPR)	EC-02
Contain biomedical laboratory science (BLS)	EC-03
Not be a gamified implementation (digital game)	EC-04

Table 1.
List of chosen criteria.

the software's teaching efficiency; (b) whether the study is specifically about basic life support with a focus on CPR; and (c) if there are details on how the software was built, with information such as platform, the tool used, as well as additional hardware.

Two researchers from this study were delegated to do the peer review, in the final choice. In addition to the complete reading of the individual selected, the researchers discussed each article, with the aim of both being in agreement on the fulfillment of the established criteria.

Figure 1, which is the PRISMA flow above, shows all stages of the study selection process until reaching the definition of the chosen works, which were 35 and are listed in **Table 2**:

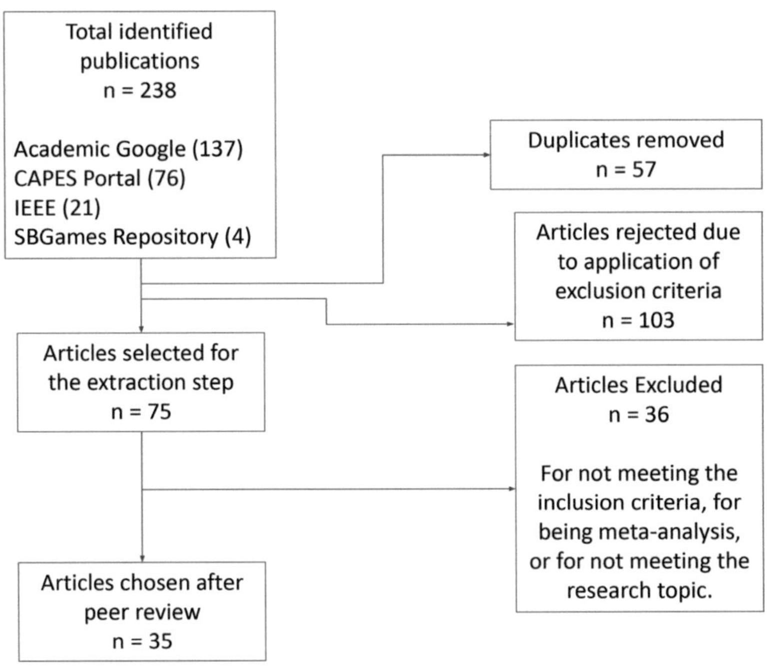

Figure 1.
Chapter selection process (prepared by the authors).

Year of publication	Amount	Published works
2021	01	A16—[17] "Comparative Effectiveness of Simulation versus Serious..."
2020	10	A07—[18] "Hands 2 Help: Educational "Serious Game" for teaching..." A08—[19] "Use of the serious game as an innovative educational strategy..." A11—[20] "Comparing Basic Life Support Serious Gaming Scores With..." A12—[21] "Developing a virtual simulation game for nursing..." A14—[22] "Precourse Preparation Using a Serious Smartphone Game on..." A19—[23] "A Pilot Study of CPR Quality Comparing an Augmented..." A20—[24] "Immersive Virtual Reality-Based Cardiopulmonary..." A22—[25] "Virtual Reality simulation technology for cardiopulmonary" A33—[26] "A Serious Game on the First-Aid Procedure in..." A34—[27] "Exploring User Needs in the Development..."
2019	10	A03—[28] "Proposal to build a serious game as an instrument for..." A05—[29] "Building a Health Game for Basic Life Support: Teaching..." A13—[30] "FASim: A 3D Serious Game for the First Aid Emergency" A15—[31] "Comparative evaluation of video-based online course versus..." A17—[32] "Virtual Reality Simulation Technology for Cardiopulmonary..." A21—[33] "Feasibility of an augmented reality cardiopulmonary..." A26—[34] "Holo-BLSD—A Holographic Tool" A29—[35] "Gamifying autonomous CPR training" A30—[36] "Comparing the effects on learning outcomes" A35—[37] "From experiencing to critical thinking: a contextual..."
2018	03	A04—[38] "Developing a Digital Game for Building First Aid Basics" A09—[39] "CPRforblind: A video game to introduce cardiopulmonary..." A31—[40] "Data analytics of mobile serious games"
2017	05	A06—[41] "Simulators for teaching cardiopulmonary resuscitation:..." A23—[42] "Affordable Hi-fidelity VR based CPR simulator" A24—[43] "Development and Evaluation of a Corrective" A28—[44] "A Serious Game For cardiopulmonary" A32—[45] "Designing an Engaging and Informative Application..."
2016	06	A01—[6] "Health in Academic Gaming Literature: an analysis..." A02—[46] "Serious Game developed in Health: Integrative Literature..." A10—[47] "Cardiopulmonary Resuscitation Training by Avatars:..." A18—[48] "Success factors for serious games to enhance learning:..." A25—[49] "Relieve A markless Assistant" A27—[50] "A Game Designed to Promote the Cardiopulmonary"

Table 2.
Number of articles per publication year.

3.5 Data extraction and synthesis

The process of filling out the protocol generated an extensive file, of more than 100 pages, the result of data collection carried out with the START tool [16], cataloging information such as title, abstract, authors, keywords, journal, year, publication link, and others comments. After the complete reading, the comprehension of the selected works is also included.

A synthesized document was built in Microsoft Excel to solidify the information of interest to this study, found in the reading of each article. These were the items: (1) if there is the identification of the application of artificial intelligence techniques (yes or no); (2) year of publication of the material; (3) platform used (PC, mobile devices, virtual or augmented reality glasses and other peripherals); (4) existence of volunteers in the study (yes or no); and (5) country that carried out the survey. There

Figure 2.
Cloud of words by the title of selected works.

was space for extra information such as type of peripheral, manufacturer, and tool/ language used to create the software.

Due to the use of a tool to support the management of systematic reviews (START), it was possible to obtain the word cloud graph according to the chosen studies, with a sampling of the words that were most present in the titles (**Figure 2**).

In the following section of results and analysis, it is possible to check the graphs that were generated with this information. Out of curiosity, it is worth mentioning that among the excluded works, even with the generated search string, there were themes a little out of contexts, such as a study on fictions such as “Boys Love” [51] and collective consciousness [52].

4. Results and analysis

When analyzing the relationship between the origin of publications and quantity, **Figure 3** shows that most of the selected works are of European origin.

Another interesting observation is the representation by region, with 11 works from Europe (Denmark, Spain, Netherlands, England, Italy, Portugal, and Sweden), six in Asia (India, Japan, Pakistan, Thailand, and Taiwan), four in North America (Canada, United States), three in the Middle East (Iran and Turkey), two in Africa (South Africa and Algeria) and, finally, one work in Oceania (Australia), respectively. Brazil is the only representative of South America, with nine works highlighted.

In **Figure 4**, which shows the number of publications per year, it is possible to note that the years 2019 and 2020 were the most productive for the topic, with a total of 20 works, eight of which for mobile devices. Perhaps because studies on games are increasingly in evidence, accompanied by the popularization of smartphones, which favors and encourages the emergence of new applications. According to an annual study by the company NewZoo, released in May this year, the number of active users on this device specifically reached the amount of 3.8 billion around the world, with a growth of 6% compared to the previous year. And a 56% share of that audience is found in the Asia-Pacific region [53].

It was possible to observe that in **Figure 5**, referring to the platforms used in the selected studies, there is almost a tie between the option of implementation on PC and on mobile devices (smartphones and tablets). One of the chances that the PC is

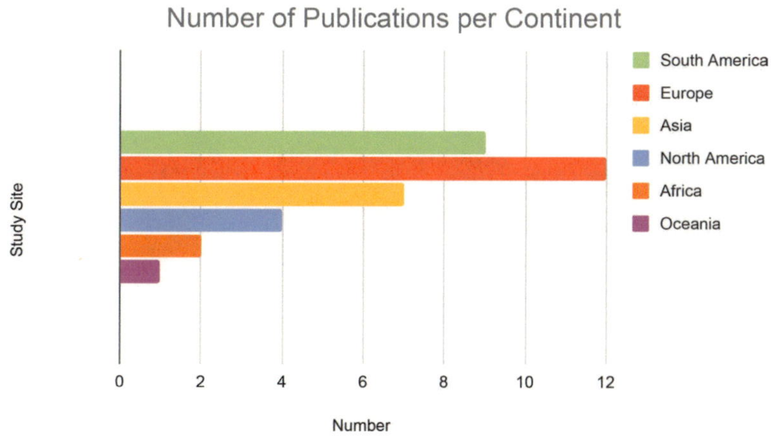

Figure 3.
Number of publications by continent.

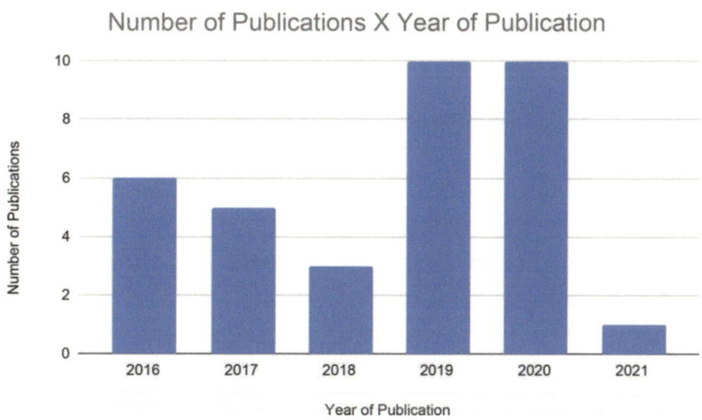

Figure 4.
Number of publications per year.

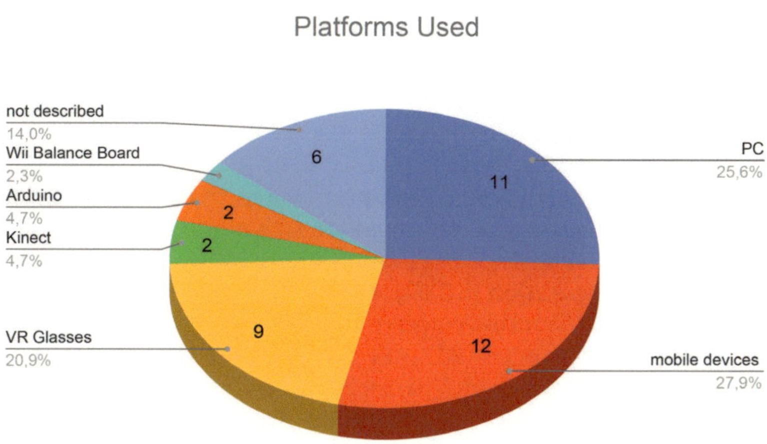

Figure 5.
Platforms used in selected studies.

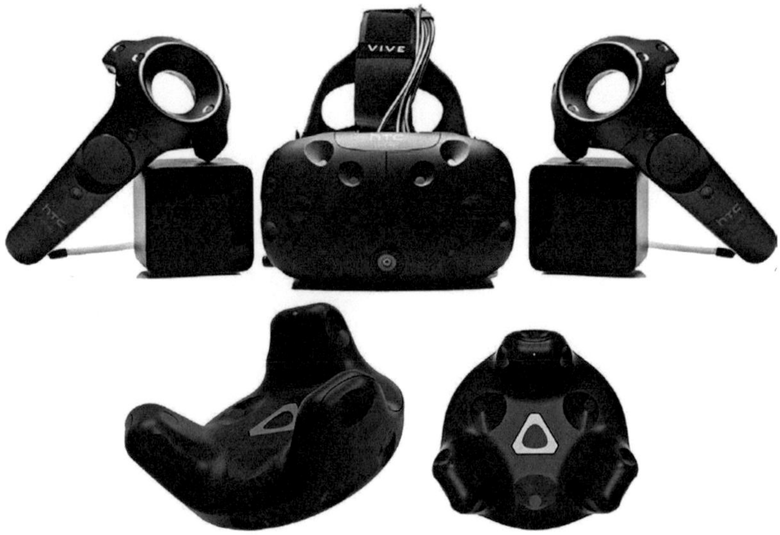

Figure 6.
HTC Vive Kit [12] with Trackers [59].

among the most chosen is due to the convenience of generating builds in the default configuration of most game development tools. Mobile devices, on the other hand, follow slightly in the lead as they are the platform with the greatest ease of access among users, even game development tools do not offer build generation by default. And according to another study by NewZoo, the global gaming market is dominated by mobile devices, with 52% of revenue [54].

In third place, there are the Virtual Reality Glasses (VR), with nine studies, representing almost 21%. They appear to be gaining in popularity, particularly in the European, Asian, and North American regions. The models that appeared in this selection were HTC Vive [55], Rift Glasses, and Go Glasses [56]. There were also three studies with the MS HoloLens [57], which makes use of mixed reality. These are modern and very interesting technologies, but they have a high market value, increasing costs, and access for the vast majority of users.

Anyway, it's worth opening a parenthesis here to detail VR devices, head-mounted displays (HMDs). The HTC Vive was used in the works A17—Canada [32] and A22—Italy [25]. Both also made use of complementary peripherals, the HTC Trackers [58], to improve the experiment's effectiveness, as shown in **Figure 6** below. Just like the Unity3D software [60] was chosen for the development of the applications.

HTC Vive has a total of 70 sensors spread out in its set (glasses and controls), allowing the user to perform fluid and high-precision movements. However, for good functioning, the device needs to be calibrated for each new user, especially if there is a difference in height. It is classified as PC VR (tethered VR) as it is dependent on a rugged computer for its operation [61, 62]. It is still possible to expand the number of sensors, by inserting HTC Trackers in the set [58].

The model used in both studies (A17—**Figure 7**/A22—**Figure 8**) was the first version from 2016, which arose from a partnership between North American Valve and Taiwanese HTC. Currently, the manufacturer already has other models on the market, such as Vive Flow, Vive Pro, Vive Focus, and Vive Cosmos [55].

Following the details of the models found in the selection of studies, we have two variants of Facebook property: the Oculus Rift, which was used in study A23—India [42];

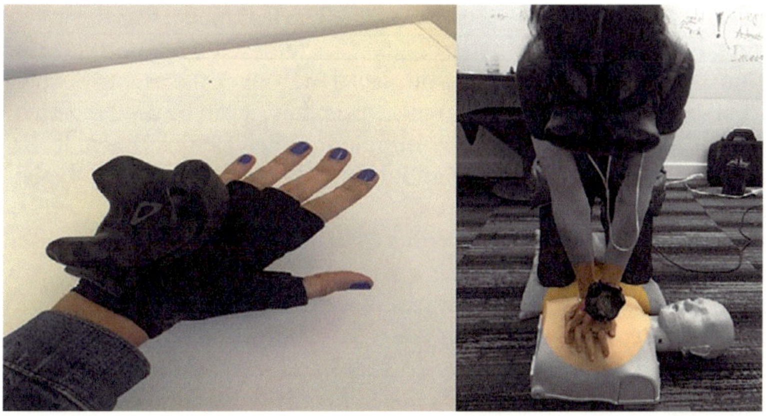

Figure 7.
HTC vive kit and trackers used on study A17 [32].

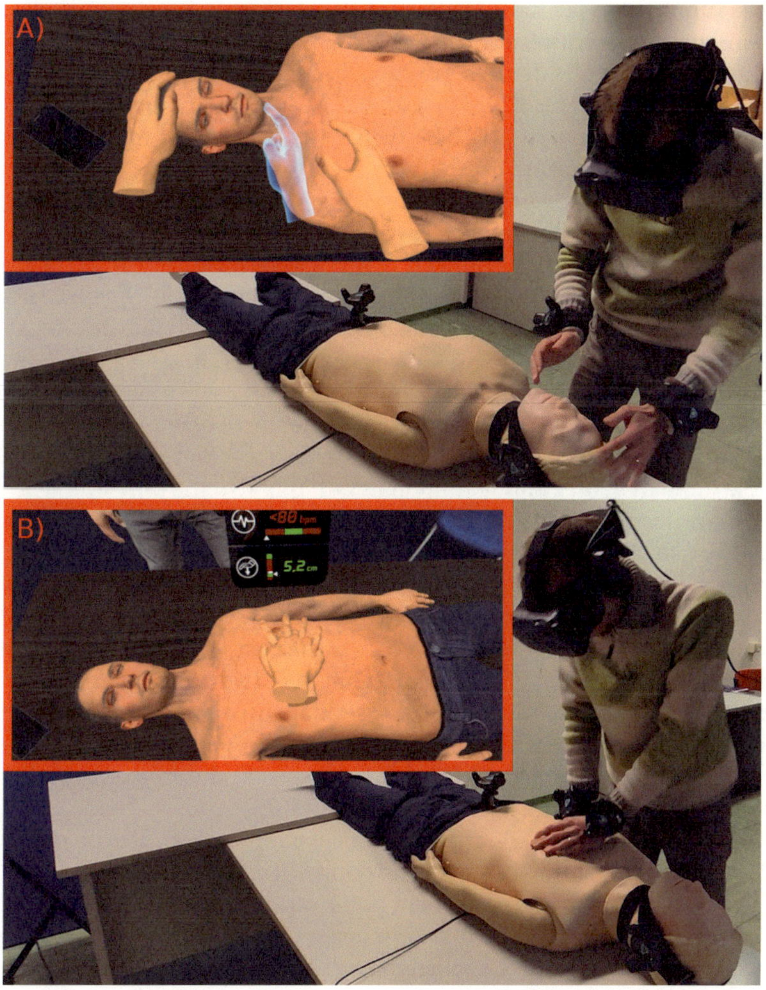

Figure 8.
Study A22 [25] uses the HTC vive kit with their software. Figure (A) is related to the measurement of the individual's carotid pulse. The figure (B) is related to cardiac compressions. Both on VR and on real life.

and the Oculus Go, in study A34—Australia [27]. The Rift model was the manufacturer's debut product, having been launched in 2016. It is a tethered VR type, comes with a sensor called Constellation, which needs to be placed in front of the user to enable tracking of the position, both sitting and sitting foot. Optionally, it can be used together with two peripherals for inserting hands in the environment, called Oculus Touch. It also has native Xbox console controller support for Windows only [63]. In study A23, only the HMD was used, with no complementary accessories.

In contrast, the other model, the Oculus Go used in study A34, is classified as an all-in-one (standalone VR) type. It does not need a powerful computer to work, as it is self-sufficient: it contains all the hardware components required to work, without the need for additional accessories, such as the aforementioned trackers [62]. But it has hand controls (Oculus Go hand controllers) for handling menus and other interactions.

The Unreal Game Engine [64] was chosen for the development of the application in the A34 study, as it allows the use of textures, has editing capacity, great customization power, in addition to providing a final product with high visual and realistic quality [27]. According to Hillman (2019), the development environment in this software has support for the most important VR platforms, with a unified set of tools that uses consistent procedures for the interaction between components [65].

Finally, among the HMDs devices, we have the Microsoft HoloLens, having been used in three of the selected studies: A19 [23] and A21 [33], both of American origin (USA); and in the European A26—Italy [34]. This device was introduced to the market in March 2016. But it acts a little differently from its competitors, making use of augmented reality (AR), by projecting for the user through the glasses on the device, information, and 3D digital graphics, overlaid on the image viewed in the real world [57].

And here comes the concept of mixed reality (MxR). He appears to be somewhat confused at first, but Bekele (2021) presents a study that details the nuances between these definitions [66]. The HoloLens is also considered a wearable device, as it allows the user to use it as a complementary accessory: as smart glasses, being called Augmented Reality Smart Glasses (ARSGs) [67].

Studies A19 and A21 only mention the use of the device, without providing further details about its application. By contrast, study A26 details his experiment, serious game modeling, and application in detail. It is also the only one who mentions the tool used for development, which was Unity3D. A common detail for the three studies was the use of a mannequin to assist in immersing the simulation for the user.

Following the analysis of peripherals used in the selected studies, Microsoft's motion sensor, Kinect [68], remains tied with Arduino, with two incidences each. Probably because of the difficulty of access and the cost involved. To use Kinect, you must have an Xbox 360 console [69] or an XOne. It is also possible to perform implementations on the PC, but it has the extra cost of acquiring the device [43, 49].

Arduino [59] is hardware with the characteristic of open source and low cost to implement. Provides great freedom for the developer. In the studies mentioned, it was used together with VR, mobile, and PC devices [24, 25].

The Nintendo Wii Balance Board [70] was used in one studies of Japanese origin. The implementation was in conjunction with MS Kinect and the PC [43]. And finally, the option without platform registration/citation, with six incidences.

Figure 9 clearly shows the number of studies that have the participation of volunteers in order to validate the experiment: 74% yes and 25% no. This was one of the criteria for selection of studies and identification of qualities, item (a) of Section 3.4 this work. Therefore, the importance of this criterion for validating the software developed for any study is noted.

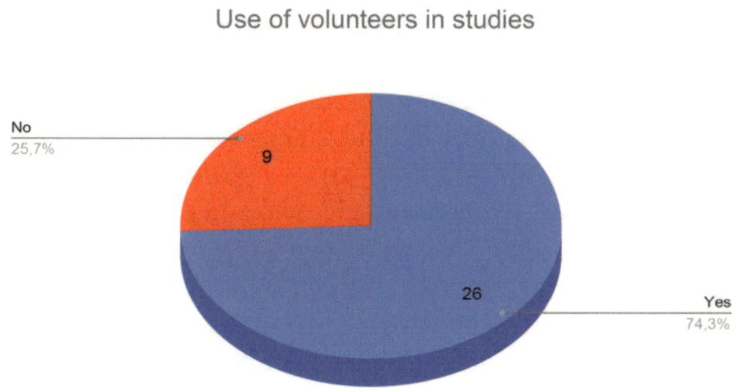

Figure 9.
Presence of volunteers in the study.

4.1 Quality assessment result

As listed in Section 3.4 (criteria for selection of studies and identification of qualities) of this chapter, three quality criteria were defined. The first refers to letter (a) if the study uses volunteers to test the efficiency of teaching the software. **Figure 6** shown in the previous item brings this answer and 26 articles make use of this guideline. The second refers to letter (b) if the study is specifically about basic life support with a focus on CPR. Yes, the 35 selected studies deal specifically with this theme, and only five of them are indirect, presenting different approaches, but still, the CPR theme is present. The third refers to letter (c) if there are details on how the software was built, with information such as platform, the tool used, as well as additional hardware. Yes, these data were included, and it is possible to check the information generated in the figures and graphs in this article.

Another interesting information raised was that eight studies [22, 24, 25, 32, 34, 44, 45, 50] explained in their texts the use of the Unity3D game engine [60], a well-known tool as a game engine and quite popular among game developers.

There were also mentions of other technologies such as MS Visual Studio [43, 71] for programming in the C# language, the use of the Japanese RPG Maker VX game engine [37, 72], and the robust Unreal Game Engine [64], software owned by Epic Games.

4.2 Critical comparisons

Some articles presented simulators as a serious gaming resource. Define the difference between simulators and serious games, what is a simulator and what is a game. If it's a simulator, is it considered a game too?

Other articles addressed the competition of learning using online videos and serious games, where the effectiveness was greater when using videos, but volunteers reinforced that learning with games was more interesting. But games are software that uses multimedia resources (audio, video, and images), there is even a genre that applies this resource in the game: the visual novels. The popularization of games focused on RCP, which is of this genre, would not be more interesting for the target audience? [73].

Twenty-nine (29) works mentioned the platform used. A tie-in incidences between the PC and mobile devices was registered. What will be the ideal platform to implement serious health games? Each platform has its own advantages and disadvantages, including in-game features and mechanics. Or would the focus be on reaching all of them? But the more cross-platform the game is, the higher production costs will be involved, which can make the creation of the software unfeasible.

It is worth mentioning that Pavkov et al. (2017) present a comparison between five game engines, aimed at the development of serious games. With the advantages and disadvantages of each, the study proposes a criterion in order to facilitate the choice of the tool by the future developer [74]. In contrast, Cowan and Kapralos (2017) argue that it is common for educators to need to hire well-skilled developers to assist in programming serious game projects [75]. The authors even recommend other game engines not identified here in this study, for those who aspire to develop games but lack programming knowledge.

Another interesting observation is that nine studies were registered that present games aimed at VR devices. It shows an interesting trend, but for developing countries with scarce financial resources for research (aka Brazil), unfortunately, this is a difficult trend to follow here. In addition to limiting the target audience too much, VR devices have a high market value compared to other platforms such as PC or smartphones, which makes production even more impossible.

On the other hand, studies that mentioned the use of Microsoft Kinect and Nintendo Balance Board devices (A24 and A25) may become meaningless, as both peripherals left the manufacturing line, no longer being marketed [76].

4.3 Existing gaps

Conducting this systematic mapping raised some interesting points to be mentioned. One was to note that there is no academically validated research regarding existing commercial implementations of serious CPR-focused games. A quick search of game app stores such as Steam for PC and Google Play for Android mobile devices reveals a plethora of health-oriented games (Heath Games). Some even had great commercial success, with millions of units sold.

It is remarkable the fact that a commercial game has a greater appeal for playfulness/entertainment, dispensing with realistic educational rigor, with concepts to be worked on and solidified. But it is an undeniable fact that they are much more popular among the general consumer public, compared to serious games, mostly with academic backgrounds.

Another interesting perception for future works would be a literary, integrative, or systematic review focused on some of the platforms identified in the research, such as only mobile devices, only consoles, VR devices, or PC.

It was also possible to notice the need for curation of games available in digital stores. In order to get a guide or perhaps a list of health game recommendations. Not just with a commercial view, but with academic validation.

Finally, it was noted the need for more systematic reviews and mappings within the area, not only focused on CPR, but on other aspects that Health Games may have examples.

5. Conclusion

The selected works were of great value for the understanding of this study, of how a systematic mapping is developed, with all its detailed and careful process that must be carried out.

Next, the answers for each research question mentioned in item 3.1 follow:

- Q1: How have serious games been used to teach CPR protocols? In carrying out this work, it was possible to verify that serious games have been used with different approaches, such as through PCs, mobile devices, and HDMs. Auxiliary peripherals (MS Kinect, Nintendo Balance Board, and Arduino) were also used to obtain better results for the user in immersing the simulations.

- Q2: How effective are serious games in teaching CPR? In the 26 studies (74.3%) that used volunteers to validate the results, it was possible to notice that the use of games for learning generates engagement, providing a deepening in thinking (deep thinking), as can be seen in the reports of the consulted students to validate the A35 job, for example. Because games have the characteristic of playfulness, being "playing" to carry out the activity, in a virtual environment controlled by rules, the involvement in the activity to achieve the best result, undoubtedly generates an increase in interest among users. And the studies that used VR as the main device have the potential to be even more immersive, due to the characteristics of the device.

- Q3: What is the state of the art of serious health-focused games for teaching CPR? It is believed that this work made a portion of contribution to the academic community in trying to potentially answer this question.

- Q4: What platforms are used to experience the games? This detail can be seen in its entirety in item 4.

To improve this work, the evaluation criteria must be revised, listing weights for each one, in order to have numerical data to define the quality of the selected articles. Review some criteria and perhaps define further details between them. Also expand the number of databases searched, for those who know how to find more jobs in the South American region.

It was also possible to notice that the area of the Serious Games focused on health is relatively new when observing the age of the relevant events in the area: SBGames will complete 20 editions by the end of this year, but only 5 years ago it has a track dedicated to the theme; SEGAH is reaching its 9th edition. This fact opens up great possibilities for studies and research that have not yet been carried out, as the field is relatively new.

It was also possible to note that there is still a lot of room for research using new technologies, such as VR glasses, Microsoft Kinect, which have only one study each.

Author details

Ingrid Nery Mendes[1*], Maicon de Araújo Nogueira[2], Filipe Valente Mendes[3], Otavio Noura Teixeira[1] and Viviane Almeida dos Santos[1]

1 Postgraduate Program in Applied Computing (PPCA), Amazonian Development Center in Engineering (NDAE), Federal University of Pará (UFPA), Tucuruí, Pará, Brazil

2 Stricto Sensu Graduate Program, Master in Health Education in the Amazon (ESA), State University of Pará (UEPA), Belém, Pará, Brazil

3 Full Degree in Artistic Education (Plastic Arts), Federal University of Pará (UFPA), Belém, Pará, Brazil

*Address all correspondence to: ingrid.mendes@tucurui.ufpa.br

References

[1] Coelho JR, Nunes AL. The importance of electronic games for the development and learning of children from a Piagetian perspective. Scientific Journal (Esfera Academica Humanas MultiVix). 2018;**2**(1):3

[2] Games for Health. Available from: https://gamesfor.health/

[3] Games for Health Europe. Available from: https://www.gamesforhealtheurope.org/

[4] Games for Change. Available from: https://www.gamesforchange.org/

[5] Games Studies. Available from: http://gamestudies.org/2004

[6] Vasconcellos MS, Carvalho FG, Capella MAM, Dias CM, Araújo IS. Health in academic gaming literature: An analysis of SBGames publications. In: Proceedings of SBGames. Culture Trail—Full Paper. 2016. Available from: http://www.sbgames.org/sbgames2016/downloads/anais/157759.pdf [Accessed: 27 April 2021]. Article A01

[7] SBGames Health Trail. Available from: https://www.sbgames.org/sbgames2021/health/

[8] IEEE SEGAH. Available from: https://www.segah.org/2021/

[9] Gonzalez MM et al. I guidelines for cardiopulmonary resuscitation and emergency cardiovascular care of the Brazilian Society of Cardiology. Arquivos Brasileiros de Cardiologia. 2013;**101**(2):3

[10] Silva TRB, Nogueira MA, Antonia MM. Knowledge of the nursing team about the care of the potential donor in brain death. Revista de Enfermagem da UFPI. 2016;**5**(4):24-30

[11] Araujo MH, Stein M, Romão JJS. Game and serious games: Concept and best practices for an analysis of Volkswagen Game—SpaceCross. In: Proceedings of SBGames—Art and Design Trail—Short Paper. 2012. Available from: http://sbgames.org/sbgames2012/proceedings/papers/artedesign/AD_Short1.pdf [Accessed: 27 April 2021]

[12] Derryberry A. Serious Games: Online Games for Learning. I'mSerious.net. White Paper. Available from: https://iktmangud.files.wordpress.com/2014/09/online-games-for-learning.pdf [Accessed: 27 April 2021]

[13] Bergeron B. Appendix A: Glossary. In: Developing Serious Games. Hingham: Charles River Media; 2006. p. 398

[14] Gorbanev Y, Agudelo-Londoño S, González RA, Courts A, Orchards A, Delgadillo V, et al. A systematic review of serious games in medical education: Quality of evidence and pedagogical strategy. Medical Education Online. 2018;**23**(1):1438718. DOI: 10.1080/10872981.2018.1438718

[15] Kitchenham B, Charters S. Guidelines for performing systematic literature reviews in software engineering. Tech. Rep. EBSE 2007-001, Keele University and Durham University Joint Report. 2007

[16] State of the Art through Systematic Review (START). Available from: http://lapes.dc.ufscar.br/tools/start_tool

[17] Farsi Z, Yazdani M, Butler S, Nezamzadeh M, Mirlashari J.

Comparative effectiveness of simulation versus serious game for nursing students in cardiopulmonary resuscitation: A randomized control trial. International Journal of Computer Games Technology. 2021;**2021**:6695077. DOI: 10.1155/2021/6695077. Article A16

[18] Alves JR, Nogueira RC. Hands 2 help: Educational serious game for teaching first aid to victims of cardio-respiratory arrest (CPA) [master's dissertation]. RJ: Vassouras University; 2020. Available from: http://mestradosaude.universidadedevassouras.edu.br/arquivos/dissertacoes/Dissertacao_Ruy_Nogueira.pdf [Accessed: 27 April 2021]. Article A07

[19] Siqueira TV, Nascimento JSG, Oliveira JLG, Regino DSG, Dalri MCB. Use of the serious game as an innovative educational strategy for learning cardiopulmonary resuscitation: An integrative review. Revista Gaúcha de Enfermagem. 2020;**41**:e20190293. DOI: 10.1590/1983-1447.2020.20190293. Article A08

[20] Aksoy ME. Comparing basic life support serious gaming scores with hands-on training platform performance scores: Pilot simulation study for basic life support training. JMIR Serious Games. 2020;**8**(4):e24166. DOI: 10.2196/24166. Article A11

[21] Keys E, Luctkar-Flude M, Tyerman J, Sears K, Woo K. Developing a virtual simulation game for nursing resuscitation education. Clinical Simulation in Nursing. 2020;**39**(C): 51-54. DOI: 10.1016/j.ecns.2019.10.009. Article A12

[22] Phungoen P, Promto S, Chanthawatthanarak S, Maneepong S, Apiratwarakul K, Kotruchin P, et al. Precourse preparation using a serious smartphone game on advanced life support knowledge and skills: Randomized controlled trial. Journal of Medical Internet Research. 2020;**22**(3):e16987. DOI: 10.2196/16987. Article A14

[23] Leary M, McGovern SK, Balian S, Abella BS, Blewer AL. A pilot study of CPR quality comparing an augmented reality application vs. the standard audio-visual feedback manikin. Frontiers in Digital Health. 2020;**2**(1):8. DOI: 10.3389/fdgth.202.001. Article A19

[24] Yang C-H, Liu S-F, Lin C-Y, Liu C-F. Immersive virtual reality-based cardiopulmonary resuscitation interactive learning support system. IEEE Access. 2020;**8**:120870-120880. DOI: 10.1109/ACCESS.2020.3006280. Article A20

[25] Buttussi F, Chittarro L, Valent F. A virtual reality methodology for cardiopulmonary resuscitation training with and without a physical mannequin. Journal of Biomedical Informatics. 2020;**111**:103590. DOI: 10.1016/j.jbi.2020.103590. Article A22

[26] Boada I, Rodriguez Benitez A, Thió-Henestrosa S, Soler J. A serious game on the first-aid procedure in choking scenarios: Design and evaluation study. JMIR Serious Games. 2020;**8**(3):e16655. DOI: 10.2196/16655. Article A33

[27] Moore N, Yoo S, Poronnik P, Brown M, Ahmadpour N. Exploring user needs in the development of a virtual reality-based advanced life support training platform: Exploratory usability study. JMIR Serious Games. 2020;**8**(3):e20797. DOI: 10.2196/20797. Article A34

[28] Oliveira GS, Almeida FL, Reis MBO, Guimarães TR. Proposal to build a serious game as an instrument for promoting health among Brazilian children and

adolescents. Time-machine doctor, awareness of preventable diseases. In: SBC—Proceedings of SBGames. Games and Health Workshop—Short Paper. 2019. Available from: https://www.sbgames.org/sbgames2019/files/papers/WorkshopJogosSaude/198475.pdf [Accessed: 27 April 2021]. Article A03

[29] Santos RR, Ferreira SC, Santos MC, Lima LG, Borges DAM. Building a health game for basic life support: Teaching strategy. Learning in health. In: Proceedings of the Seminar Technologies Applied to Education and Health. 2019. Available from: https://www.revistas.uneb.br/index.php/staes/article/view/8208 [Accessed: 27 April 2021]. Article A05

[30] Benkhedda S, Bendella F. FASim: A 3D serious game for the first aid emergency. Simulation & Gaming. 2019;**50**(6):1-21. DOI: 10.1177/1046878119865913. Article A13

[31] Sena DP, Fabricio DD, Silva VD, Bodanese LC, Franco A. Comparative evaluation of video-based online course versus serious game for training medical students in cardiopulmonary resuscitation: A randomized trial. PLoS One. 2019;**14**(4):e0214722. DOI: 10.1371/journal.pone.0214722. Article A15

[32] Almousa O, Prates J, Yeslam N, Mac Gregor D, Zhang J, Phan V, et al. Virtual reality simulation technology for cardiopulmonary resuscitation training: An innovative hybrid system with haptic feedback. Simulation & Gaming. 2019;**50**(1):6-22. DOI: 10.1177/1046878118820905. Article A17

[33] Leary M, McGovern SK, Balian S, Abella BS, Blewer AL. Feasibility of an augmented reality cardiopulmonary resuscitation training system for health care providers. Heliyon. 2019;**5**(8):e02205. DOI: 10.1109/ACCESS.2020.3006280. Article A21

[34] Strada F, Bottino A, Lamberti F, Mormando G, Ingrassia PL. Holo-BLSD—A holographic tool for self-training and self-evaluation of emergency response skills. In: IEEE Transactions on Emerging Topics in Computing. 2019. DOI:10.1109/TETC.2019.2925777 [Accessed: 27 April 2021]. Article A26

[35] Santos I, Sa-Couto C, Vieira-Marques P. Gamification in CPR—A review of game dynamics and mechanics. In: 2019 14th Iberian Conference on Information Systems and Technologies (CISTI). DOI:10.23919/cisti.2019.8760603 [Accessed: 27 April 2021]. Article A29

[36] Aksoy E. Comparing the effects on learning outcomes of tablet-based and virtual reality-based serious gaming modules for basic life support training: Randomized trial. JMIR Serious Games. 2019;**7**(2):e13442. DOI: 10.2196/13442. Article A30

[37] Chang C-Y, Kao C-H, Hwang G-J, Lin F-H. From experiencing critical thinking: A contextual game-based learning approach to improving nursing students' performance in electrocardiogram training. Educational Technology Research and Development. 2020;**68**:1225-1245. DOI: 10.1007/s11423-019-09723-x. Article A35

[38] Bishop, GP, Carneiro N, Sarinho VT. Developing a digital game for building first aid basics. In: Proceedings of SBGames. 3rd Workshop on Games and Health—Short Paper. 2018. Available from: https://www.sbgames.org/sbgames2018/files/papers/WorkshopJogosSaude/186836.pdf [Accessed: 27 April 2021]. Article A04

[39] Rodríguez-Benitez A, Boada I, Thió-Henestrosa S, Sbert M. CPRforblind: A video game to introduce cardiopulmonary resuscitation protocol to blind people. British Journal of Educational Technology. 2018;**49**(4): 636-645. DOI: 10.1111/bjet.12627. Article A09

[40] Lukosch H, Cunningham S. Data analytics of mobile serious games: Applying Bayesian data analysis methods. International Journal of Serious Games. 2018;**5**(1):19-36. DOI: 10.17083/ijsg.v5i1.222. Article A31

[41] Reis AV, Rocha RN. Simulators for teaching cardiopulmonary resuscitation: Analysis of CardioSIM as a Brazilian solution. In: Proceedings of SBGames—Art & Design Trail—Short Paper. 2017. Available from: http://www.sbgames.org/sbgames2017/papers/ArtesDesignShort/174967.pdf [Accessed: 27 April 2021]. Article A06

[42] Durai SIV, Arjunan R, Manivannan M. Affordable hi-fidelity VR based CPR simulator with haptics feedback. In: 2017 IEEE International Symposium on Haptic, Audio and Visual Environments and Games (HAVE). 2017. DOI:10.1109/have.2017.8240349 [Accessed: 27 April 2021]. Article A23

[43] Higashi E, Fukagawa K, Kasimura R, Kanamori Y, Minazuki A, Hayashi H. Development and evaluation of a corrective feedback system using an augmented reality for the high-quality cardiopulmonary resuscitation training. In: 2017 IEEE International Conference on Systems, Man, and Cybernetics (SMC). 2017. DOI:10.1109/smc.2017.8122692 [Accessed: 27 April 2021]. Article A24

[44] Latif MH, Ajmal M, Ahmad F, Alam J, Saleem A. LA-VIE: A serious game for cardiopulmonary resuscitation. In: 2017 IEEE 5th International Conference on Serious Games and Applications for Health (SeGAH). DOI:10.1109/segah.2017.7939297 [Accessed: 27 April 2021]. Article A28

[45] Foldager N, Hansen H, Tewes MS, Bjørner T. Designing an engaging and informative application about first aid: Gamification and humor as design elements in a serious game. In: Gaggi O, Manzoni P, Palazzi C, Bujari A, Marquez-Barja J, editors. Smart Objects and Technologies for Social Good. GOODTECHS 2016. Lecture Notes from the Institute for Computer Sciences, Social Informatics and Telecommunications Engineering. Vol. 195. Cham: Springer; 2017. DOI: 10.1007/978-3-319-61949-1_9. Article A32

[46] Deguirmendjian SC, Miranda FM, Mascarenhas SHZ. Serious game developed in health: Integrative literature review. The Journal of Health Informatics. 2016;**8**(3):110-116. Article A02

[47] Creutzfeldt J, Hedman L, Felländer-Tsai L. Cardiopulmonary resuscitation training by avatars: A qualitative study of medical students' experiences using a multiplayer virtual world. JMIR Serious Games. 2016;**4**(2):e22. DOI: 10.2196/games.6448. Article A10

[48] Ravyse WS, Suagnet Blignaut A, Leendertz V, Woolner A. Success factors for serious games to enhance learning: A systematic review. Virtual Reality. 2016;**21**(1):31-58. DOI: 10.1007/s10055-016-0298-4. Article A18

[49] Loconsole C, Frisoli A, Semeraro F, Stroppa F, Mastronicola N, Filippeschi A, et al. RELIVE: A markerless assistant for CPR training. IEEE Transactions on Human-Machine Systems. 2016;**46**(5):755-760. DOI: 10.1109/thms.2016.2586756. Article A25

[50] Boada I, Rodríguez-Benitez A, Garcia-Gonzalez JM, Thió-Henestrosa S, Sbert M. 30 : 2: A game designed to promote the cardiopulmonary resuscitation protocol. International Journal of Computer Games Technology. 2016;**2016**:8251461. DOI: 10.1155/2016/8251461. Article A27

[51] Zhang C. Loving boys twice as much: Chinese women's paradoxical fandom of boys love fiction. Women's Studies in Communication. 2016;**39**(3):249-267. DOI: 10.1080/07491409.2016.1190806

[52] Bourazeri A, Pitt J, Arnab S. Enabling collective awareness of energy use via a social serious game. EAI Endorsed Transactions on Serious Games. 2017;**4**(13):e2. DOI: 10.4108/eai.12-27-2017.153510

[53] 2021 Global Smartphones Users. Active Smartphone Users per Region. Global Mobile Market Report. May 2021. Available from: https://newzoo.com/wp-content/uploads/2016/04/Newzoo_2019_Smartphone_Users_per_Region2-1024x576.png [Accessed: 31 July 2021]

[54] 2021 Global Games Market. Per Device & Segment with Year-on-Year Growth Rates. Global Games Market Report. June 2021. Available from: https://newzoo.com/wp-content/uploads/2016/03/Newzoo_Global_Games_Market_by_Segment_June2021-1024x576.png [Accessed: 31 July 2021]

[55] HTC VIVE. Available from: https://www.vive.com/us/

[56] Oculus Rift, Go and Quest. Available from: https://www.oculus.com/

[57] Microsoft HoloLens. Available from: https://www.microsoft.com/en-us/hololens

[58] HTC Trackers. Available from: https://www.vive.com/us/accessory/tracker3/

[59] Arduino. Available from: https://www.arduino.cc/

[60] Unity3D. Available from: https://unity3d.com/en/get-unity/download

[61] Alves A, et al. Chapter 4: Developing applications in virtual reality with HTC VIVE in unity C#. In: 19th Symposium on Virtual and Augmented Reality (SVR 2017). 2017. pp. 67-81

[62] Angelov V, Petkov E, Shipkovenski G, Kalushkov T. Modern virtual reality headsets. In: 2020 International Congress on Human-Computer Interaction, Optimization and Robotic Applications (HORA). 2020. DOI:10.1109/hour49412.2020.9152604

[63] Nafees A. Oculus Rift: A Rift in Reality. ResearchGate Publication; 2016. DOI: 10.13140/RG.2.2.33985.61281

[64] Unreal Game Engine. Available from: https://www.unrealengine.com/en-US/

[65] Hillmann C. Unreal for VR: Basic ingredients for an oculus go project setup. In: Unreal for Mobile and Standalone VR. Berkeley, CA: Apress; 2019. pp. 87-140. DOI: 10.1007/978-1-4842-4360-2_4

[66] Bekele MK. Mixed reality: A bridge or a fusion between two worlds? In: Champion EM, editor. Virtual Heritage: A Guide. London: Ubiquity Press; 2021. pp. 93-103

[67] Kalantari M, Rauschnabel P. Exploring the early adopters of augmented reality smart glasses: The case

of Microsoft HoloLens. In: Augmented Reality and Virtual Reality (Progress in IS). Springer; 2018. pp. 229-245. DOI: 10.1007/978-3-319-64027-3_16

[68] Microsoft Kinect. Available from: https://www.xbox.com/zh-CN/consoles/xbox-one/accessories/kinect

[69] Microsoft XBox. Available from: https://www.xbox.com/pt-BR/games/xbox-360

[70] Nintendo Wii Balance Board. Available from: https://www.nintendoblast.com.br/2011/06/plug-andblast-balance-board-wii.html

[71] Visual Studio. Available from: https://visualstudio.microsoft.com/en-us/

[72] RPGMaker VX. Available from: https://www.rpgmakerweb.com/products/rpg-maker-vx

[73] Andrew J, Henry S, Yudhisthira AN, Yulyani A, Syarifah DP. Analyzing the factors that influence learning experience through game based learning using visual novel game for learning Pancasila. In: 4th International Conference on Computer Science and Computational Intelligence 2019 (ICCSCI). 2019. DOI:10.1016/j.procs.2019.08.177 [Accessed: 31 July 2021]

[74] Pavkov S, Frankovic I, Hoic-Bozic N. Comparison of game engines for serious games. In: 2017 40th International Convention on Information and Communication Technology, Electronics and Microelectronics (MIPRO). 2017. DOI:10.23919/mipro.2017.7973518

[75] Cowan B, Kapralos B. An overview of serious game engines and frameworks. In: Intelligent Systems Reference Library. 2017. pp. 15-38. DOI:10.1007/978-3-319-49879-9_2

[76] Good O. Kinect is officially dead. Really. Officially. It's dead. Polygon; 2017. Available from: https://www.polygon.com/2017/10/25/16543192/kinectdiscontinuedmicrosoft-announcement

Chapter 5

3D Computer Graphics and Virtual Reality

Branislav Sobota and Miriama Mattová

Abstract

This chapter is dedicated to the description of 3D computer graphics used for the needs of virtual reality. Virtual reality (VR) is the use of computer technology to create a 3D virtual environment. The chapter presents some graphical features used in an environment as well as an explanation of good design practice. The chapter contains also the description of lighting settings, 3D objects/models and virtualization sequence, camera, and scenes where the wheelchair simulator is used as an example of the implementation environment.

Keywords: 3D graphics, 3D objects, virtual reality, web-based collaborative environments, lighting settings, object geometry, texturing

1. Introduction

Technology is advancing in the directions in which it eases life. In the transport industry, it helps to get from point A to point B faster. In the education industry, it helps to obtain knowledge in a more effective way. In the health industry, it helps to operate or diagnose easier. Those are very few examples of how technology is simplifying life. The process of technology advancing is defined in a few steps. The first one is the curiosity about how things work. The second is the willingness to learn. The third is understanding and the fourth is experimenting. The creation of 3D computer graphics had a similar process. Firstly, there was a curiosity about how people see and percept the world. Secondly, it was a willingness to learn how the visual information from the outside world is getting into person. From that point, people realized that eyes were the ones to blame. Willingness to learn continued in this area and later on there was an understanding. The understanding of how the eyes work helped to create the first photographic camera. Lastly, there was experimenting in this field which led to the first photographs. As it is the nature of humans, we wanted to achieve more. We wanted to make those photographs to be animated and displayed somewhere. The development of this technology has gained rapid momentum. Soon, people were able to watch a short movie on canvas. With an introduction of the first computer, those two areas soonly merged. Before that, there was a need to represent data calculated on computer. Soon, it was represented on a simple display. As computers advanced and were more complex, there was a need to represent calculated data more effectively. Computer monitors were introduced and computer graphic was born. Before it reached the possibility of 3D displaying, 2D scenes were presented. The most popular one was television. This is the greatest example so-called eye (visual) and computer content. As people realized how effective and entertaining computers can be, they started to wonder if we can simulate our world. Information from the computer area, eyes (visual) information and mathematics combined, and 3D computer graphics was born.

3D graphics opened a new world of experimenting in the area. As we live in a three-dimensional world, it gives us another level of possibilities of how to simulate and improve the virtual world-virtual reality. Representation of virtual reality via 3D technologies gives more authenticity and credibility. Virtual reality (VR) and related technologies are penetrating almost every aspect of our lives [1]. Nowadays, it is one of the most developing technologies that could significantly increase the level of interactivity [2]. This is greatly used in every industry. It has a great impact on health industry as it is possible to create projects like wheelchair simulators for people who are temporarily or permanently immobilized. Those people are provided with a number of courses to gain wheelchair skills but it is often financially unavailable. Virtual reality was proven as an effective tool for disabled people [3].

This chapter is dedicated to the description of 3D computer graphics used for the needs of virtual reality. The wheelchair simulator is used as an example of the implementation environment. Wheelchair simulator was built using web technologies via A-frame framework [4]. It is a simulation of a collaborative environment in VR (developed in authors' home laboratory LIRKIS at Department of Computers and Informatics, Technical University of Košice) where an immobilized patient should learn using an electric wheelchair. Guest can join the same 3D virtual environment and watch patient deal with the obstacles. Whole simulator contains multiple scenes with different obstacles. The pilot project was also implemented in a unique 3D virtual LIRKIS CAVE environment (**Figure 1**).

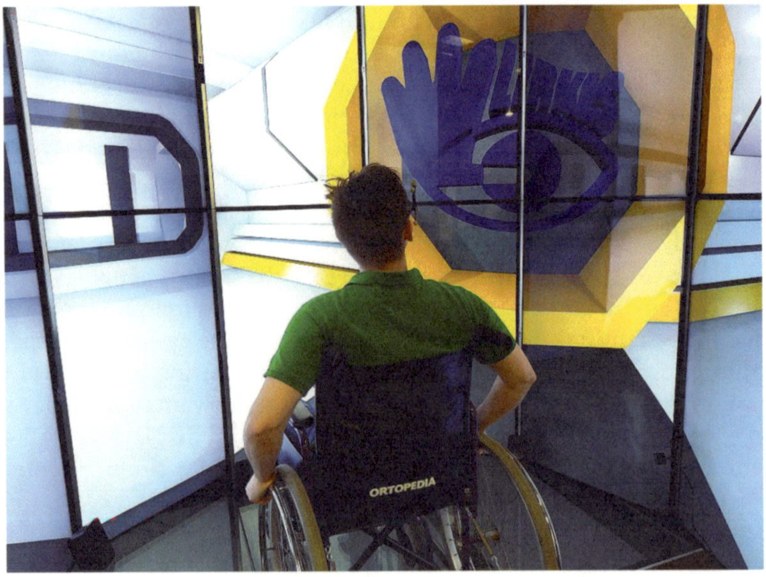

Figure 1.
Patient on wheelchair in 3D virtual LIRKIS CAVE environment.

DOI: http://dx.doi.org/10.5772/intechopen.102744

2. Related work

There are many works related to 3D computer graphics in virtual reality. Some of the works focus on different virtualization algorithms, others try a new methodology of improving realistic look while maintaining the smoothness of application. Related work below simply focuses on new methodologies or approaches to 3D computer graphics or comparing them.

Chen in [5] describes the rapid creation of photorealistic virtual reality content with consumer depth cameras. Work focuses on the demonstration of a complete end-to-end pipeline for the capture, processing, and rendering of view-dependent 3D models in virtual reality from a single consumer-grade RGB-D camera. The result of this pipeline is a 3D mesh with view-dependent textures suitable for real-time rendering in virtual reality. Progressive feedback point cloud rendering for virtual reality display is the aim of [6]. Work presents a novel approach to a progressive feedback-driven rendering algorithm. This algorithm uses reprojections of past views to accelerate the reconstruction of the current view. The presented method is tested against previous methods, showing improvements in both rendering quality and interactivity. In the [7] is indicated virtual content creation using dynamic omnidirectional texture synthesis. This work proposes optimization that synthesizes a high resolution view-dependent texture map for any virtual camera location. Synthetic textures are generated by uniformly sampling a spherical virtual camera set surrounding the virtual object, thereby enabling efficient real-time rendering for all potential viewing directions.

The paper [8] describes three views of virtual reality: non-immersive virtual reality, and it discusses the advantages and applications of non-immersive VR systems. Immersive and non-immersive VR systems are compared and hybrid possibilities are reviewed.

Kautz et al. [9] present new algorithms and techniques for the acquisition and real-time interaction with complex textured 3D objects and shows how these results can be seamlessly integrated with previous work into a single framework for the acquisition, processing, and interactive display of high-quality 3D models. In addition to pure geometry, such algorithms also have to take into account the texture of an object and its reflectance behavior. The research [10] presents a novel 3D user interface for the immersive design review. This research focuses on the development of a novel immersive user interface, a smart 3D disk with a set of widgets. During the immersive sessions, the user can activate functionalities using cost-effective pointing devices and the conceived 3dUIs projected into the more complex CAD functionalities such as surfaces shape modification.

3. 3D graphic in VR

3D computer graphics are graphics that use a three-dimensional representation of geometric data that are stored in a computer for the purpose of performing calculations and rendering 2D images. Those geometric data are not graphics until they are presented on a screen. Geometric data can be perceived as objects or models. To build up a 3D world in virtual reality, it is needed to include graphic elements such as:

- *Lighting*: Lighting settings give the light in the scene, manage shadows, light sources as well as their intensity.

- *3D objects*: 3D objects or models are a base to build up a scene. It defines, what the scene will contain in the largest spectrum.

- *Camera*: Camera manages the process of visibility, culling, and rendering of 3D objects on the screen.

- *Scenes*: Scene is a space where all elements above are put. It also defines the size of space in a virtual world.

A combination of all mentioned graphic elements can result in scene as shown in **Figure 2** left. Another 3D scene can be seen in drone simulation environment represented in **Figure 2** right.

3.1 Rendering

Rendering can be perceived as the final process of creating the final 2D image from the prepared scene. These images are played at some speed rate. It is called frames per second (FPS). This variable indicates how many final 2D images are rendered in 1 second on the screen. The higher the value, the smoother is playback image. FPS can be used as final control of how great was handled graphics optimalization with a given graphics processing unit (GPU). It is important to handle draw calls effectively as it results in smoother applications. Rendering can be divided into real-time rendering and non-real-time rendering.

Real-time rendering is used in interactive media such as games and simulations. It means final 2D images are being processed in real time, according to where the camera is positioned and how it is rotated in the scene. In real life, movement of a person looking around the room can be random as he/she is moving head is not defined speed or rotation. The same logic applies to a camera in media. Therefore, there is no algorithm to predict the exact movement of camera, as it is random input from user, for final 2D images to be calculated beforehand. Using this method, it is needed to keep in mind that human eyes need at least 24 frames per second rendered for a successful illusion of movement. In VR headsets, frames per second must be quite higher, otherwise, it is possible to notice black spots when rotating the head. Black spots around the edges can cause the loss of the illusion of being in a 3D environment.

Figure 2.
Example representation of the wheelchair simulator scene and the drone simulation scene.

Non-real-time rendering is used for such media in which it is possible to predict the sequence of images. These are non-interactive media such as film or video. It can take much more time to render in order to obtain higher image quality.

Speed of rendering depends not only on effective optimalization of application but also on hardware specification of device like GPU, CPU, and memory, for example. This is a reason why the same application can be smooth in desktop computers but not in VR standalone headsets. Therefore, there is a need for 3D graphics knowledge and hardware specification knowledge to optimize application if it is built for multiple platforms.

3.2 General optimalization

General optimalization in 3D applications consists of multiple steps. But firstly, it is important to realize what is the desired result. To call application optimalized, it is desired to achieve a count of draw calls as small as possible, limit the unnecessary calculations or optimize complexity. This can be achieved through several steps, for example:

- *Light optimalization*: Uneffective placement, settings, and types of lights sources can greatly impact calculations during rendering which results in lower FPS. Next Section 4 focuses on these parameters as well as implemented lights in wheelchair simulator.

- *Simplifying objects*: It includes creating lots of details (LOD), having an object with no unnecessary polygons, and using adequate resolution of textures (Section 5).

- *Camera settings*: Since camera view is the final view of what user will see, it is important to pass to the renderer only information that needs to be handled in real time and to prevent all the unnecessary information to be calculated (Section 6).

- *Scene logic*: If it is not necessary for application logic, scenes should be divided into smaller scenes. A wide scene with open field can cause computational load, therefore it is recommended to reconsider scene logic. Section 7 focuses on scene design in wheelchair simulators.

4. Lighting setting

Lights in 3D virtual environments are necessary for objects to be visible. The same logic applies in real world. Without light, it is not possible to see your surroundings. There are several types of lights, but most common are *directional light*, *point light*, *spotlight*, and *ambient light*. The difference between these lights is ways of lighting up a specific area, where a light source and lighting direction is located. Lighting itself works basically the same. Light has defined its source, direction, intensity, and area, then this parameter defines the vectors which are colliding and bouncing from the objects in dependence on how close the light source is, how strong light intensity is, and area of light up space. Wheelchair simulator uses ambient light in all scenes. **Figure 3** represents scenes with ambient lightning.

One of the scenes is designed to be at night. Scene is called crosswalk-night. As is mentioned above, ambient light is everywhere and it illuminates every object evenly from every side. **Figure 4** shows the scene.

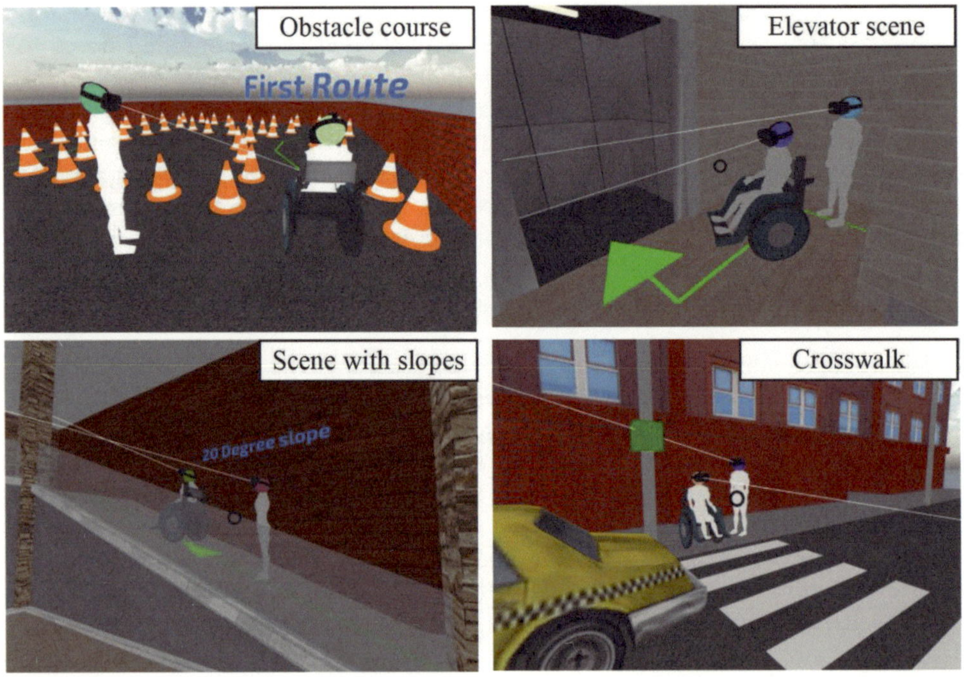

Figure 3.
Four out of five scenes with ambient lighting.

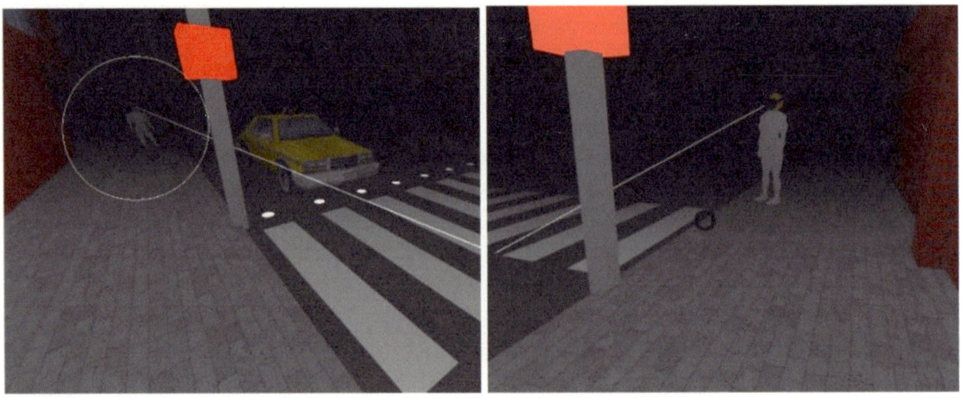

Figure 4.
Representation of crosswalk-night scene.

It is possible to observe that besides ambient light, fog is implemented as well. Fog allows to shade the objects at a certain distance in a predetermined color. It is implemented on camera of the user and not as a light object. Therefore, fog is not perceived as a lighting setting but as a shading setting. In this case, distance of shading is few meters from the user, and color of shading is black. This way it is possible to simulate first-person view at night, samely as in real life. From **Figure 4**, it is also possible to see the red traffic light, which signals that patient cannot cross the crosswalk as cars are moving. The red color on a box may seem a lot lighter than surroundings. It is because of emissive intensity of material on a traffic light was set on a higher value to make it seems lit. This way, it is not necessary to add the point light in the environment and it will lighten the computational core. Details are explained in Section 3.3. But it is necessary to keep in mind, if priority is to develop application with high rate of frames per second, it can affect the realistic look of the scene. Especially, for virtual headsets as their computational cores are not so advanced yet. **Figure 5** shows the traffic light with an implementation of point light in red color and with shadows settings.

4.1 Shaders

Thanks to specific light definition, it is possible to observe illuminated and unlit parts of an object. Light has an exact border where the illumination of object ends. Therefore, it is needed to shade the transition between illuminated and unlit parts to make it appear more natural. Shader will make a transition by adjusting the color of the nearest pixels between these two sections. **Figure 6** shows the illuminated part, shaded part, and unlit part of the cone illuminated by point light only.

Figure 5.
Traffic light with point light in red color and shadow settings.

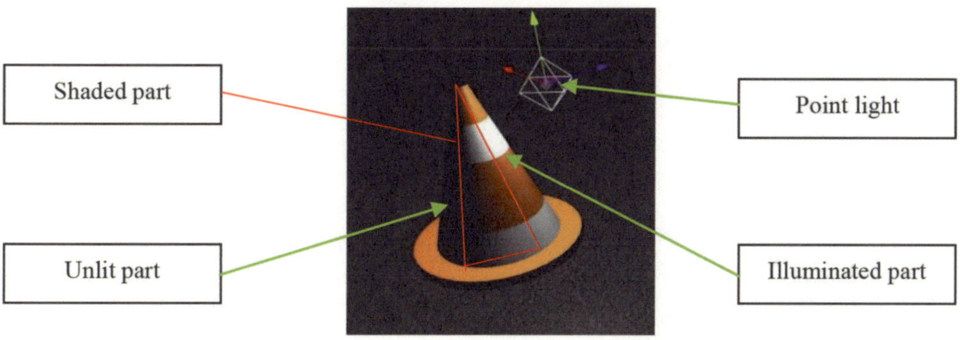

Figure 6.
Shader represented on a cone with a point light.

4.2 Lights combinations and their complexity

All mentioned light types can be used at a same time and multiple times in a scene. **Figure 7** shows implementation of spot light, which is red, point light which is blue, and directional light which is white.

All lights presented in **Figure 7** are casting shadows. All objects are casting and receiving shadows as well. This is an example of how light can simulate the real-world behavior of lighting. Even though it looks natural and realistic, it is needed to keep in mind that there is a lot of calculation done in the background. It is possible to observe that blue light is illuminating the shadowed part on a terrain from the red light as well as the not shadowed part. Directional light illuminates area of red light and blue light at once but also creates shadows on a left from the cones. All lights are penetrating areas of other light, therefore, renderer needs to consider priority of which light it will display. It is possible to bake these lights at the same time, but calculation complexity will greatly improve. Vectors from lights are crossing paths, which mean they are affecting each other. Therefore, they need to be recalculated to give a result similar, if not same, to a real world. Simply said, the more lights are used in a scene and the more they are penetrating into areas of other lights, the higher complexity lighting setting will have. If renderer needs to take shadows into account, it needs to calculate position, rotation, and size of an object for every light separately as well. And lastly, renderer needs to compare position of the object which is receiving the shadows to an object which is casting the shadow. When creating a 3D scene, it is advised to consider lighting logistics depending on the strength of the computational core as it can cause computational load which can result in small FPS.

Wheelchair simulators were using few point lights for better authenticity. But after profiling of performance was done, FPS was lower than 60 in virtual headsets. The main goal of simulator was to teach disabled person using the wheelchair and overcame obstacles. To teach that, realistic graphics were on a lower priority than smoothness of application. Due to this fact, all unnecessary lights were removed from the scene and ambient light was implemented. Ambient light is considered as the light with the

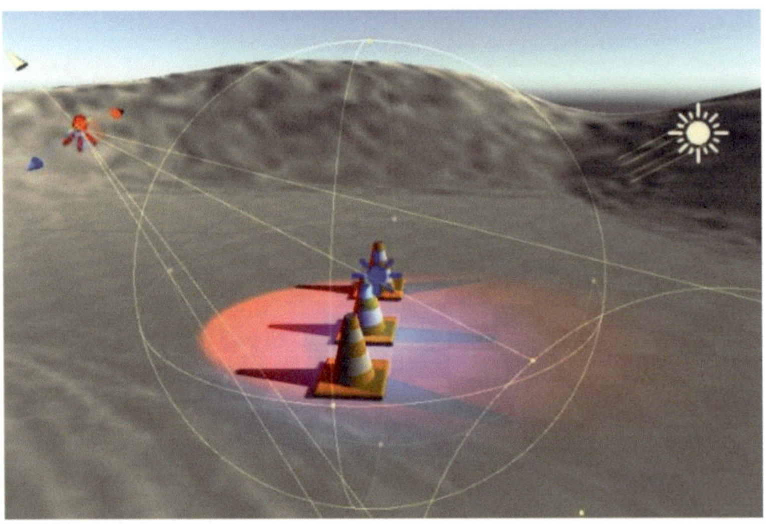

Figure 7.
Combination of spot light, point light, and directional light in a scene.

lowest cost for the computational core. Therefore, shadows were not implemented at all, as ambient light cannot cast them. If shadows cannot be cast, there is no need for a shader. Result is a simple lighting setting but a lot of computational space is saved.

5. 3D objects and virtualization sequence

The whole process of object processing and virtualization is called the virtualization sequence. A simplified 3D virtualization sequence scheme is shown in **Figure 8**. Of course, each of these steps contains a number of other sub-steps naturally. Many sub-sequences of this sequence including 3D objects modeling and editing are already standardly used in practice, and they have steady conventions and standards. The virtualization sequence is important in defining and creating VR-based systems. In principle, it determines all the virtual world objects, their way of use, and of course, the semantics (meaning) of individual objects [11].

As mentioned in the chapter, 3D computer modeling is the creation or modification of a 3D graphical virtual model using a specific software tool—3D modeling application (e.g. Blender, Autodesk 3D Studio Max, Maya, Sketchup, etc.). The modeling process of preparing/editing graphical data is similar to sculpture creation. 3D objects modeling is not an exception in this case and it is based on a four-step principle: model/object preparation, polygonal modeling, texture preparation, and texture application and filtering. The properties of these steps depend on the model/object type. The created model/object can be used for visualization, simulation, or for 3D printing.

Each model consists of *points*, *edges*, and *faces*. These are primitives that create geometry of any model. **Figure 9** represents geometry of wheelchair using primitives.

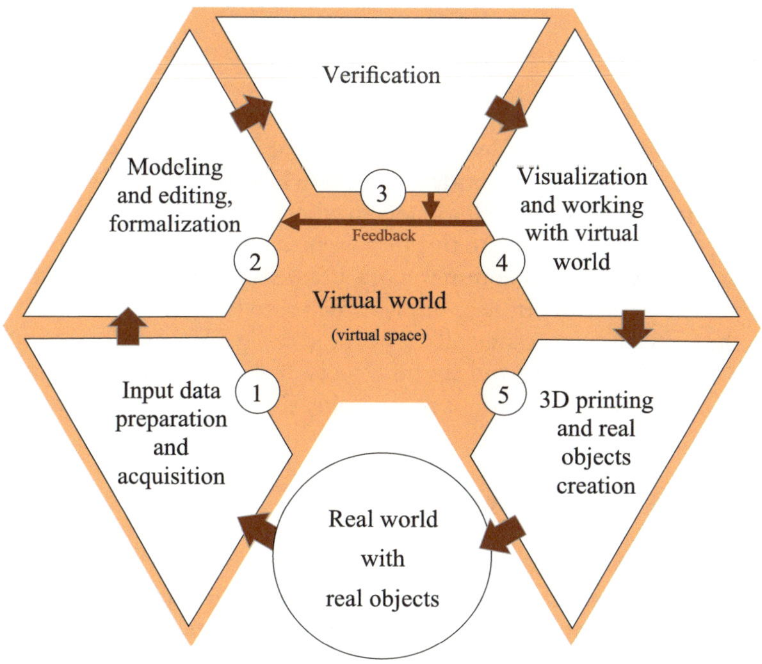

Figure 8.
A simplified 3D virtualization sequence scheme.

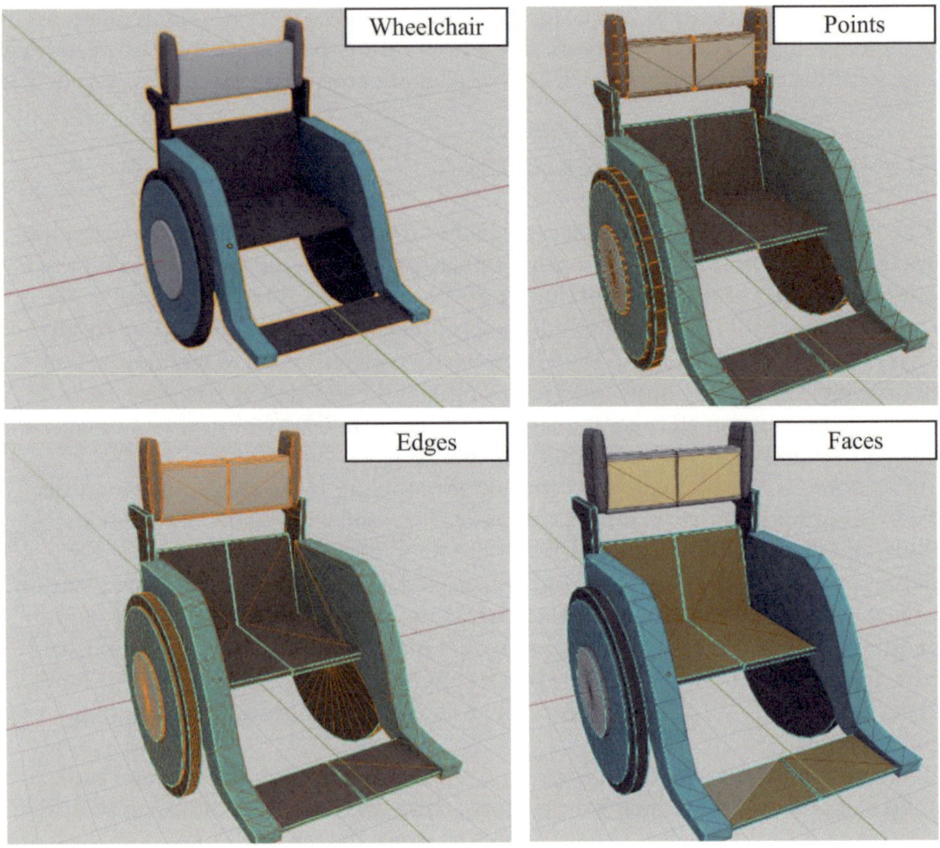

Figure 9.
Points, edges, and faces represented on cone model.

Wheelchair model has assigned base color texture. A number of polygons are in a count necessary to shape the model into a wheelchair look. Although it does not look realistic, it brings a possibility to render such model faster. Therefore, its optimal design solution for VR headsets. All models in wheelchair simulators are designed with low polygons principle. Due to this fact, scenes have united design with fast rendering. As helmets are used commercially, it is generally known that realistic look does not affect the possibility of user being drawn into the environment. The user is drawn into the environment, especially, by united design of models and their texture, interactivity with environment and sound effects. If these elements are designed creatively and are not exaggerated, user can enjoy the virtual world samely as environment with realistic look.

5.1 Objects geometry in the simulator

Target devices for wheelchair simulators are virtual headsets, therefore, it is needed to keep objects geometry as simple as possible. The higher is complexity of an object, the more it is compelling for the renderer to process. Every polygon in the object will burden renderer to calculate it and display it on the final screen. Due to this fact, many applications for virtual reality use low polygon technique of modeling. **Figure 10** shows geometry of some objects used in the simulator.

As **Figure 10** proves, models are build up the way just to represent its necessary shapes, to figure out meaning of the object. Disadvantage of low polygon objects is that neglected folds result in a non-realistic look. But it still keeps the shape to figure out the context. One of the advantages is, that this type of modeling practice does not require creating lots of details—LOD for every object. LOD is also an unnecessary feature in wheelchair simulators as scenes are not that wide for LOD to be applied. It is possible to decimate some polygons on a wheelchair or the car model, for example, those parts of model that are not so visible by a user. But a number of polygons will not change dramatically if it is desired to keep the look so it is an unnecessary step. Such a step would be worth considering if these objects were placed into environment multiple times. Then the polygon reduced will be the number of reduced polygons times the number of duplicated models.

5.2 Textures

Wheelchair simulator does not use special texturing features, such as normal mapping or Dot 3 bump mapping. It uses only simple base color textures and image textures. If the quality of the image texture is restricted to smaller size and resolution, rendering will be optimalized to desired preferences. Wheelchair simulator is using

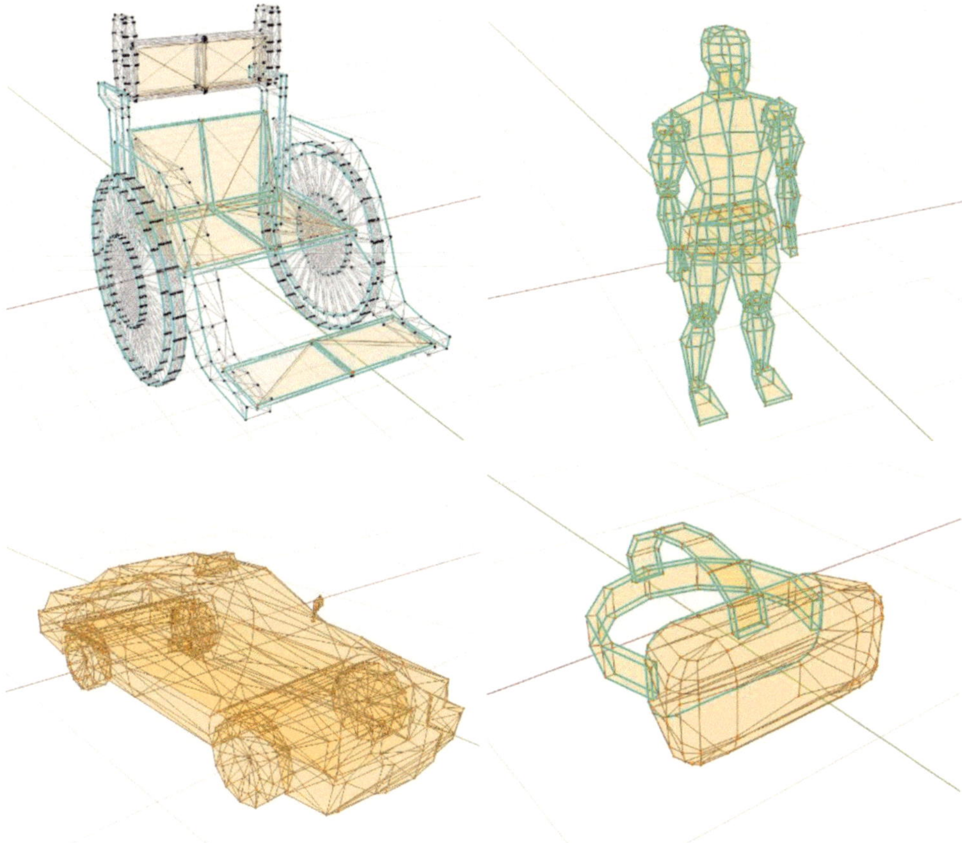

Figure 10.
Geometry representation of wheelchair, avatar, headset, and car models.

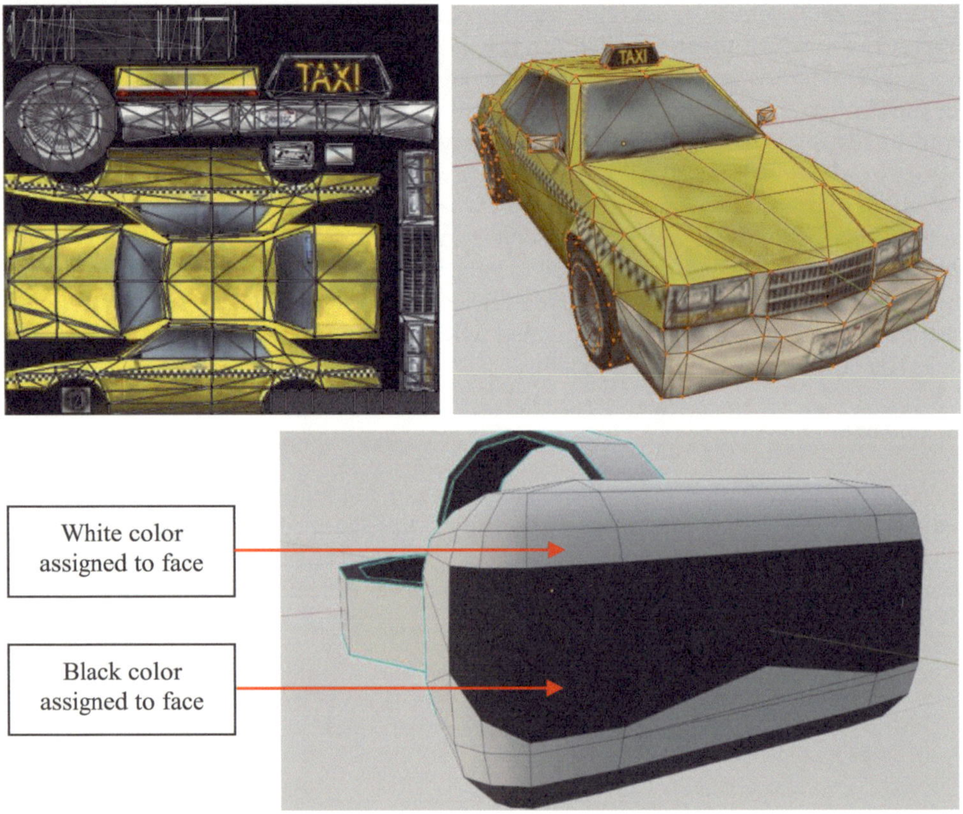

Figure 11.
Texture representation on taxi car and VR headset models.

resolution and size up to 5 megabytes. But the average sizes of image textures are 300–500 kb. Following **Figure 11** represents an example of image texture of car with a size of 387 kb and virtual headset object with assigned based colors. Image texture of a car also shows the mapping of the taxi car object. Mapping is represented with points primitives of an object. Their position defines the position of the texture on the object.

Resolution of a texture is important to be optimalized. If the texture has for example resolution of 4000 × 4000 pixels, but texture coloring and details can fit into resolution 500 × 500 pixels without any pixel compressing, it is highly recommended to change the resolution of the texture. Every time, when an object is in a field of view, piece of texture is rendered in dependence on the face shape and size of an object. Therefore, if the simple cube with 6 walls-6 faces have assigned full-size texture with resolution 4000 × 4000 pixels on each face, and top side, right side, and left side of a cube are in a field of view, the texture will be rendered 3 times with defined resolution.

6. Screen-camera

Camera and screen settings can facilitate render calculation as well. To do so, it is needed to reconsider, for example, if it is more effective to use *single pass* or *multipass*

Figure 12.
Night implementation in wheelchair simulator.

rendering. *Object visibility* based on camera distance is one of the features to consider as well. For example, occlusion *culling* is a camera feature that defines if objects should be rendered when they are behind another object. There are many effective rendering techniques that help lower the run time cost.

Besides saving computational cost, camera has some features to manipulate the graphics and visualization in scenes as well. One of the features was presented in Section 3. And in **Figure 3**. As it was mentioned before, scene called crosswalk-night is situated at night. In real world, when a person is not in a completely dark environment, he/she can see objects within few meters radius. Visibility, colors, and sharpness of the objects depend on the intensity of light. To simulate such environment, scene contains ambient light of certain intensity—but lower than the ambient light in other scenes. This will secure objects to be visible. To implement logic of seeing things in some radius around a user, a feature called fog need to be applied. Fog has a parameter, which defines what color fog will be. To simulate the night, color needs to be black. Other colors, such as red for example, will be seen in a distance as ambient light is applied all around the scene. Result is shown in **Figure 12**.

Figure 12 shows, as user goes further from the crosswalk, other objects will gradually shade as they are not in a radius of fog-free area. It can be observed, that taxi car on an upper part of the figure is visible, but as user went further back, taxi car is not visible anymore.

Others performance improvement techniques such as single pass, multi-pass, occlusion culling are predefined in used framework—A-frame.

7. Scenes-spaces

When creating a 3D application, it is good to know whether the size of the scene will be wide or small in advance. Wider scenes are used mostly when creating an outdoor environment. In this case, wheelchair simulator is using small scenes. Simulator has a scene where a disabled person needs to go around cones or cross the crosswalk. This activity is placed outside, but scenes were designed with a wall limitation on terrain edges. It is easier to solve graphic limitations with a smaller scene as it does not contain many objects. Polygons' number is lower which result in faster rendering. Following **Figure 13** presents the scenes from the distance together with the top view.

7.1 Scene with slopes

One of the wheelchair simulator's goals was to teach a disabled person to go through elevated terrain. Such obstacles are mostly situated inside buildings in real life. Mostly staircases, which have barrier-free platform placed on them, have the steepest ascent. Therefore, scene with slopes was created. Whole scene consists of one big room with two slopes on the edges of the room and patient can go through

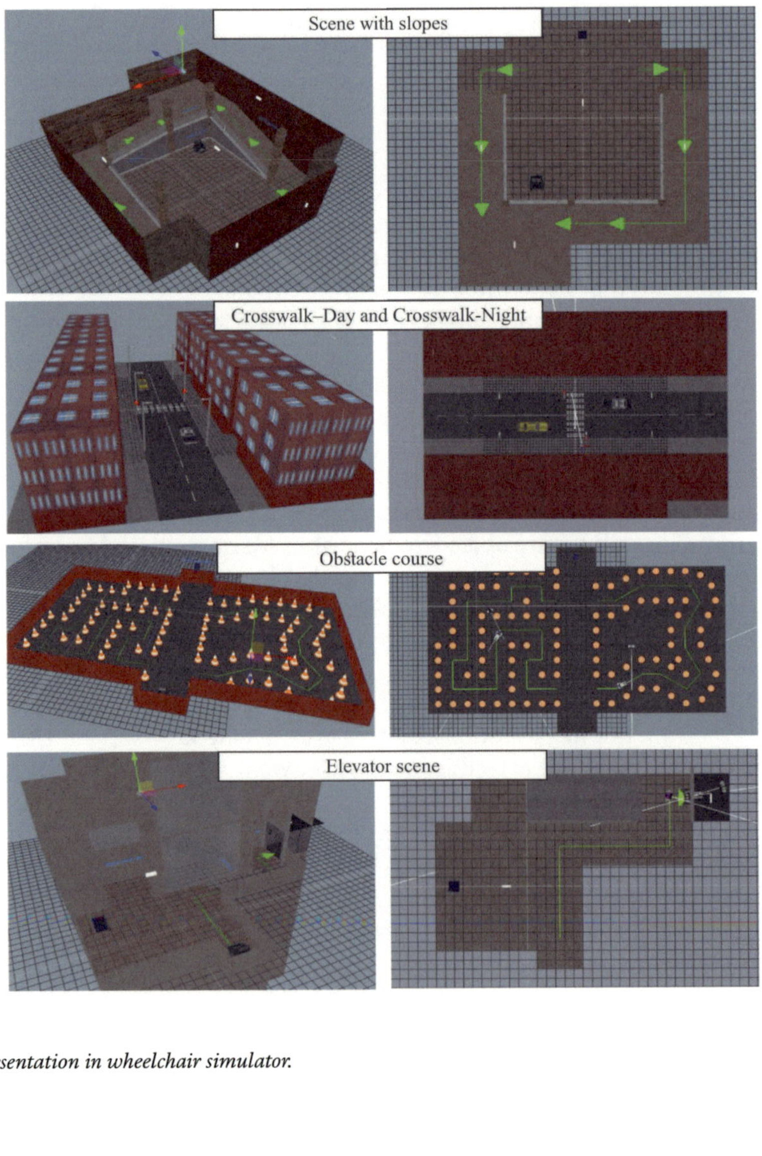

Figure 13.
Scenes representation in wheelchair simulator.

them. Scene is categorized as a small scene because its visible width is a few meters. **Figure 14** shows scene details as well as its top view with explanation.

7.2 Crosswalk-day and crosswalk-night

People living near roads will encounter crosswalks almost on daily basis. It is an unwritten rule, that crosswalk needs to be crossed as quickly as possible. Disabled person has a speed limitation. Due to this fact happens, some of them do not have confidence doing activities where they or other objects should move fast. Scene with crosswalk was designed for a person to gain confidence in a wheelchair. Scene was designed even in night mode as visibility of objects, such as cars, is limited. It is considered a small scene, even though it is situated outside because terrain is a size of a few meters. **Figure 15** shows scene details as well as its top view with explanation.

Scenario of crossing is simple. Cars are moving at a certain speed when traffic light is red. The traffic light will, after a few seconds, lights up in orange, signaling that a person can prepare for crossing. Then it lights up green signaling that person can cross the road. Cars will stop in front of the crosswalk. They do not have a collider to make sure, that if the patient is still on the road, and traffic light will light up red, cars will not move him/her out of the map. This can cause discomfort and dizziness.

7.3 Obstacle course

A person in a wheelchair must learn to control it in a first place. To learn the lever manipulation, how it affects the wheelchair and how to rotate the wheelchair, the scene with cones was created. Obstacle course scene has two routes, where a disabled person can practice its wheelchair skills. That way he/she will gain agility to continue

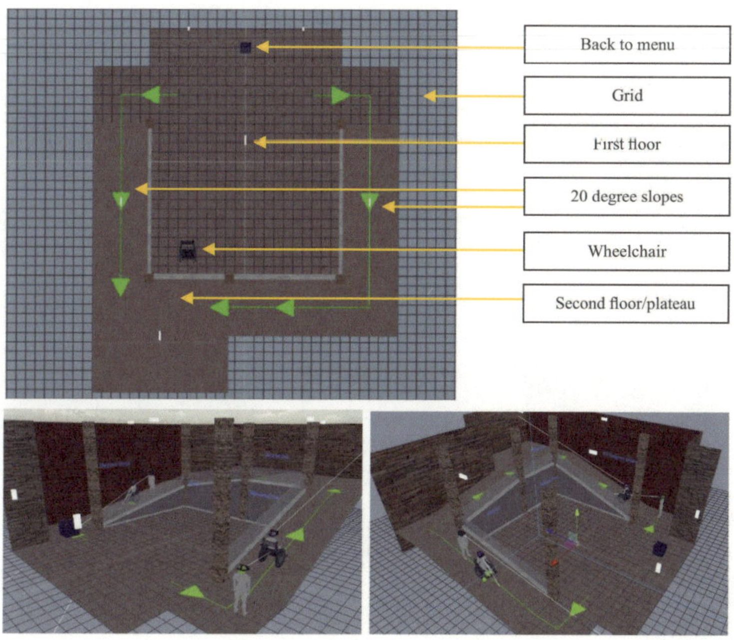

Figure 14.
Scene with slopes.

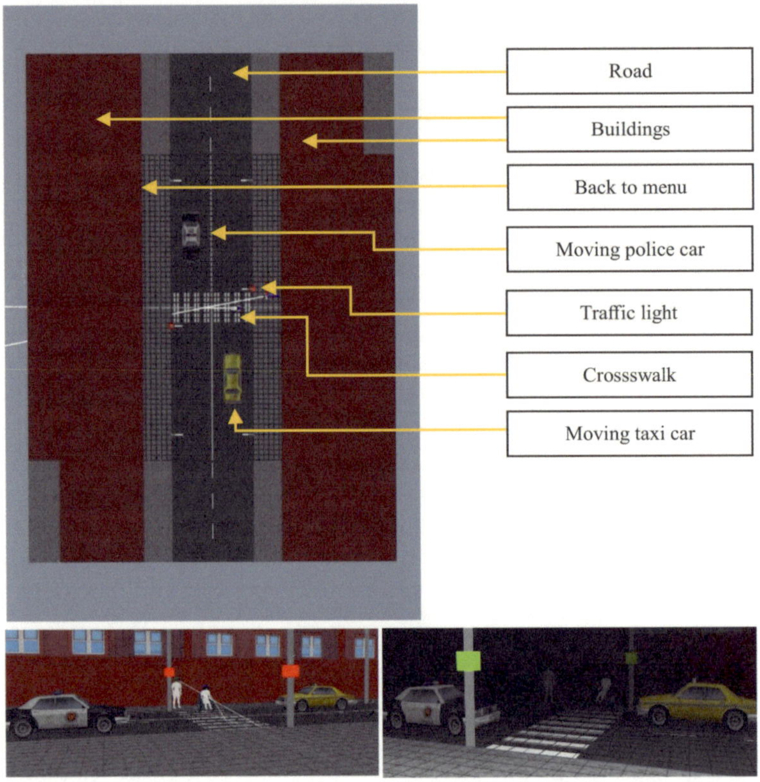

Figure 15.
Scene crosswalk-day and crosswalk-night.

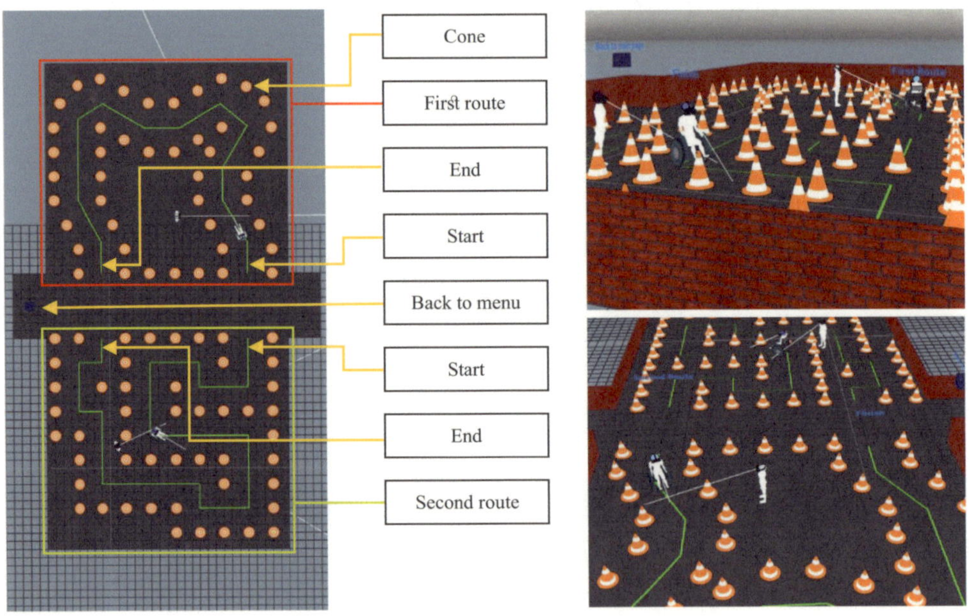

Figure 16.
Obstacle course scene.

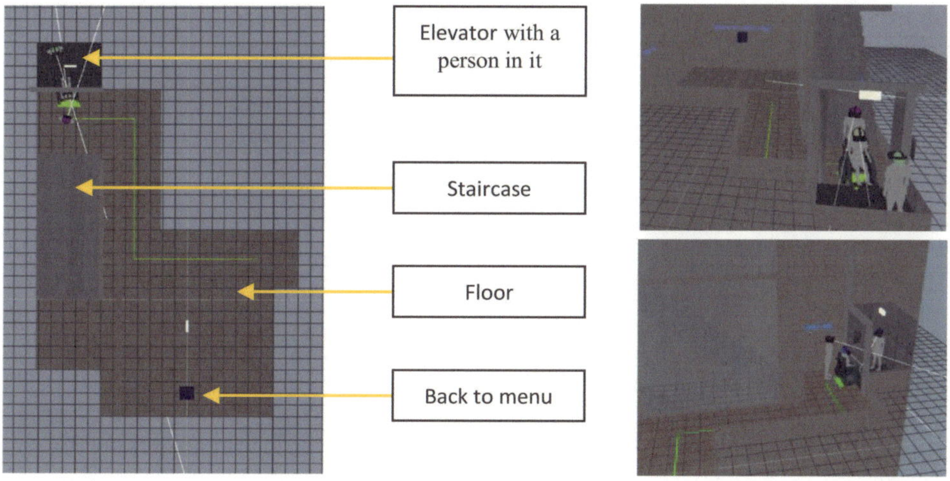

Figure 17.
Elevator scene.

with more difficult obstacles. Cones are static and have colliders so the person in wheelchair does not move them if he/she will bump into them. Same as other scenes, obstacle course is considered as a small scene. Following **Figure 16** shows scene details as well as its top view with explanation.

7.4 Elevator scene

The elevator scene was designed to teach the person in a wheelchair to go through a narrow hallway or learn how to fit in an elevator. Scene is situated in a building with a staircase. It is impossible to go through the staircase as it has a collider. Therefore, proposed scenario is that person in a wheelchair wants to go upstairs. **Figure 17** shows scene details as well as its top view with explanation.

8. Conclusion

3D computer graphics gave the possibility to create a 3D world with scenarios, where immobilized persons can learn to handle the most common obstacles. System is supported for virtual headsets, therefore, immobilized persons can experience more authentic environment. Chapter represents 3D computer graphics used in an environment together with an explanation of why they were implemented that way. Virtual reality is a technology that supports such environments. As VR technology is relatively young and it is still evolving, it was needed to implement such 3D graphics, which will not compromise the smoothness of the simulator. Therefore, simulator is implemented with prioritization of smoothness over a realistic look. Chapter also focuses on such 3D graphics combinations to give the user the best experience.

Acknowledgements

This work has been supported by the Operational Programme Integrated Infrastructure in a frame of the project: Intelligent systems for UAV real-time operation and data processing, code ITMS2014+: 313011 V422 and co-financed by the European Regional Development Fund and also it has been supported by the KEGA grant No. 048TUKE-4/2022: "Collaborative virtual reality technologies in the educational process".

Author details

Branislav Sobota* and Miriama Mattová
Technical University of Košice, Košice, Slovakia

*Address all correspondence to: branislav.sobota@tuke.sk

References

[1] Sobota B, Korečko Š, Hrozek F. On building an object-oriented parallel virtual reality system. Central European Journal of Computer Science. 2012;**2**(3):261-271

[2] Hudák M, Korečko Š, Sobota B. Enhancing team interaction and cross-platform access in web-based collaborative virtual environments. In: Proceedings of 2019 IEEE 15th International Scientific Conference on Informatics. IEEE; 2019. pp. 160-116

[3] de Ipina JML, et al. Virtual reality: A tool for the disabled people labour integration. In: Proceeding of Challenges for Assistive Technology. IOS Press; 2007. pp. 141-145

[4] Boguščiak J. Wheelchair simulator in web virtual reality [diploma thesis]. Košice: TU; 2021. p. 62

[5] Chen C, Bolas M, Rosenberg ES. Rapid creation of photorealistic virtual reality content with consumer depth cameras. In: 2017 IEEE Virtual Reality (VR). 2017. pp. 473-474. DOI:10.1109/VR.2017.7892385

[6] Tredinnick R, Broecker M, Ponto K. Progressive feedback point cloud rendering for virtual reality display. In: 2016 IEEE Virtual Reality (VR). 2016. pp. 301-302. DOI:10.1109/VR.2016.7504773

[7] Chen C, Rosenberg ES. Virtual content creation using dynamic omnidirectional texture synthesis. In: 2018 IEEE Conference on Virtual Reality and 3D User Interfaces (VR). 2018. pp. 1-2. DOI:10.1109/VR.2018.8446410

[8] Robertson GG, Card SK, Mackinlay JD. Three views of virtual reality: Nonimmersive virtual reality. Computer. 1993;**26**(2):81. DOI: 10.1109/2.192002

[9] Kautz J, Lensch HPA, Goesele M, Lang J, Seidel H. Modeling the world: The virtualization pipeline. In: Proceedings 12th International Conference on Image Analysis and Processing. 2003. pp. 166-174. DOI:10.1109/ICIAP.2003.1234044

[10] Martini A, et al. A novel 3D user interface for the immersive design review. In: 2015 IEEE Symposium on 3D User Interfaces (3DUI). 2015. pp. 175-176. DOI:10.1109/3DUI.2015.7131757

[11] Hrozek F, Korečko Š, Sobota B. Solutions for time estimation of tactile 3D models creation process. In: Advanced Machine Learning Technologies and Applications: Second International Conference (AMLTA 2014). Cairo: Springer International Publishing; 2014. pp. 498-505

Chapter 6

View Synthesis Tool for VR Immersive Video

Sarah Fachada, Daniele Bonatto, Mehrdad Teratani and Gauthier Lafruit

Abstract

This chapter addresses the view synthesis of natural scenes in virtual reality (VR) using depth image-based rendering (DIBR). This method reaches photorealistic results as it directly warps photos to obtain the output, avoiding the need to photograph every possible viewpoint or to make a 3D reconstruction of a scene followed by a ray-tracing rendering. An overview of the DIBR approach and frequently encountered challenges (disocclusion and ghosting artifacts, multi-view blending, handling of non-Lambertian objects) are described. Such technology finds applications in VR immersive displays and holography. Finally, a comprehensive manual of the Reference View Synthesis software (RVS), an open-source tool tested on open datasets and recognized by the MPEG-I standardization activities (where "I" refers to "immersive") is described for hands-on practicing.

Keywords: DIBR, RVS, view synthesis, depth map, virtual reality, rendering, 3D geometry, light field, non-Lambertian

1. Introduction

Photography has a vast history as it is used to preserve our lives' most important memories. As such, it tries to conserve a scene as realistically as possible. During the years, it evolved from the camera obscura [1, 2] where scenes were captured only for a brief moment, to black and white photography, requiring to stay still in front of the camera for long hours, to nowadays imaging devices, where the picture is captured instantaneously, digitalized, and the colors are close to what our eyes perceive [3].

Though it preserves the content of the scene, the immersion is lost, as well as the depth information, since the camera projects the scene from 3D to 2D.

To increase the immersion, the next step is to recreate the parallax of the scene, giving the opportunity to the viewer to move freely and see different perspectives, exactly as if the subject was miniaturized in front of our eyes, or the environment virtually rendered around us. Despite this desire, no device capable of acquiring the scene in its entirety directly in 3D has been designed so far.

Creating the parallax effect assumes capturing the scene from all the possible viewpoints and selecting the viewpoint to display on demand for the user's viewing position. This is physically impossible; instead, we may synthesize any viewpoint from only a couple of captured viewpoints generating all missing information following some basic assumptions [4, 5].

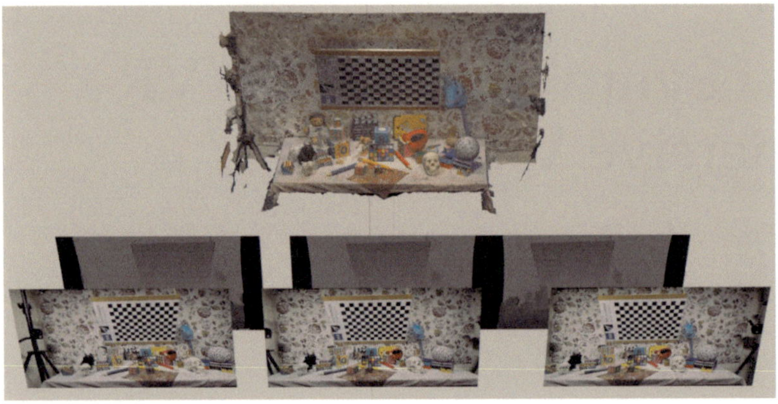

Figure 1.
DIBR brings photographs to 3D by using depth information to create new viewpoints. It preserves photorealism and allows the user to experience motion parallax.

There exist many approaches to generate novel viewpoints from input views. Early methods were based on 3D reconstruction [6, 7] to render the obtained 3D model. More recently, neural radiance fields (NeRF) [8] used machine learning methods to recreate a volumetric representation of the scene. Other methods avoid the explicit 3D information reconstruction, such as depth image-based rendering (DIBR) [9] that will be described in this chapter, or multiplane images [10, 11]. Finally, novel viewpoints can be synthesized by an intelligent interpolation using physical invariants (the epipolar plane image), rather than interpolating directly the image's colors. Representatives of this last category are the shearlet transform [12] and techniques using deep learning [13].

This chapter provides comprehensive elements to bring photographs of a natural scene to the third dimension, for example, making the captured scene immersive, through holography or virtual reality (VR). The presented 3D rendering technique differs from traditional computer graphics by its input—instead of modeling 3D objects with their geometry and materials that interact with light sources, we use photographs that are warped to follow the viewer gaze direction using the reference view synthesis (RVS) [14–16] software that follows the view synthesis process of **Figure 1**.

RVS has been developed during the exploration and standardization activities of MPEG-I – where "I" refers to Immersive – focusing on developing new compression and file formats for immersive video.

The chapter is structured as follows—the first part explains the principles of depth image-based rendering, gives an overview of the possible artifacts that can be encountered when creating a DIBR implementation, and finally, implementation details of RVS are described. The second part provides practical advice for using RVS on some example datasets.

2. Principles of depth image-based rendering

To recreate the parallax effect, we use the depth image-based rendering [9] method. It warps or distorts the input color image as a function of its associated depth map, which itself stores, for every pixel, the distance between the camera and the projected point along the camera optical axis. This method is based on the observation that a stereoscopic pair of images, for example, taken with a few centimeters shift between each other, carry the depth information of the photographed subject. As shown in **Figure 2**, the relative shift d, aka. the disparity, of foreground objects is larger than for background objects.

2.1 Projection equation and disparity

Let us consider two pinhole cameras facing an object at distance D (see **Figure 2**). The projection of this object on each image will have a disparity d. Using the similar triangles ratios of $\frac{f}{D}$ and $\frac{d_1+d_2}{B}$, we obtain:

$$d = \frac{B \times f}{D} \tag{1}$$

where B is the baseline, i.e., the distance between the two camera centers, and f their focal length.

This implies that, given two images and their depth maps, we can create a virtual view in the middle, between the inputs, by shifting the pixels over half their disparity.

Eq. 1 can be generalized to any camera settings using the pose (translation and rotation – extrinsic parameters, Eq. 2) and internal camera parameters (focal length f_x and f_y expressed in pixels in the x and y directions, and principal point $\left(pp_x, pp_y\right)$ – intrinsic parameters, Eq. (3).

We call an input image and its camera parameters an *input* viewpoint. We aim to recreate a new virtual view with given new parameters, called *target* viewpoint. For this, we deproject (i.e., from 2D to 3D) the pixels of the input image to 3D, and reproject (i.e., from 3D to 2D) them to the target image using the projection equation.

Let $P = (R|t)$ be the inverse (i.e., world to camera) 3×4 pose matrix of a camera with R the rotation matrix and t the translation:

$$P = (R\,|\,t) = \begin{pmatrix} r_{11} & r_{12} & r_{13} & t_x \\ r_{21} & r_{22} & r_{23} & t_y \\ r_{31} & r_{32} & r_{33} & t_z \end{pmatrix}, \tag{2}$$

and K its 3×3 intrinsic matrix:

$$K = \begin{pmatrix} f_x & 0 & pp_x \\ 0 & f_y & pp_y \\ 0 & 0 & 1 \end{pmatrix}. \tag{3}$$

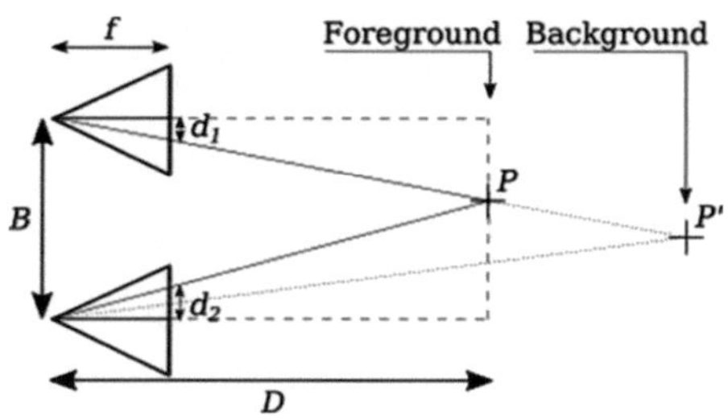

Figure 2.
Disparity $d = d_2 - d_1$ between two pixels representing the same projected point.

In homogeneous coordinates, a point $X = (x, y, z, 1)^t$ at depth D from the input camera $p_{in} = (u, v, 1)^t$ is projected to a pixel $p_{in} = (u, v, 1)^t$ following the projection Equation [6]:

$$Dp_{in} = K_{in}P_{in}X \tag{4}$$

Hence, given the input image and the depth value of the pixel, we can deproject X:

$$X = D\left(R_{in}|-R_{in}^{-1}t_{in}\right)^T K_{in}^{-1}p_{in} \tag{5}$$

Eventually, this allows to reproject X in the new camera, using Eq. 4 with P_{out} and K_{out}:

$$p_{out} \propto DK_{out}P_{out}\left(R_{in}|-R_{in}^{-1}t_{in}\right)^T K_{in}^{-1}p_{in} \tag{6}$$

To obtain the pixel value, we divide the obtained vector by the third coordinate (i.e., the depth of the point in the new camera).

Applying this operation to every pixel of the input image creates a novel view.

The core principle of DIBR is to apply this deprojection and reprojection to all the pixels of the input images, using a depth map (i.e., a single-channel image encoding the depth value of each pixel). RVS uses this basic principle, but of course, there are many pitfalls one should handle correctly. This is further explained in the following sections.

2.2 Frequent artifacts

We now know the basic principles of DIBR. Unfortunately, simply shifting the pixels of an input image in the function of their depth does not create a photo-realistic result.

The first problem is occlusion handling. When an object is visible in the input image but hidden by an object lying more in the foreground in the target, it is occluded and its pixels should not appear in the rendered image. This can be solved by choosing, among all the pixels from various objects ending up in the same pixel on the screen, the pixel with the minimal depth. A more critical problem is the one of disocclusions, for example, when an object should be visible in the target image but does not appear in the input image because it is hidden (**Figure 3a**). In that case, a

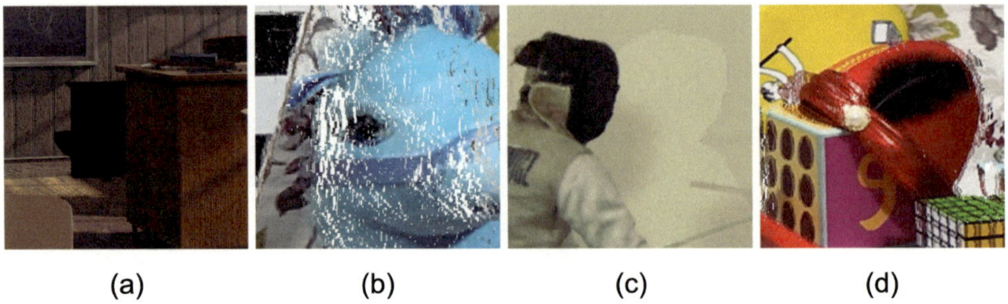

Figure 3.
(a) Disocclusion artifact (classroom dataset), (b) crack artifacts (Toystable dataset), (c) Artifacts due to inconsistency in color among the input images. (dataset fencing, courtesy of Poznan University of Technology), (d) ghosting artifacts (dataset Toystable).

hole is created in the rendered image. One solution is to add more input images in the hope to obtain this missing information [15, 17]. Another approach is to inpaint the empty pixels [18, 19]. In RVS, it is possible to choose any number of viewpoints and a basic inpainting fills the remaining disocclusions.

Cracks and dilation are other frequent DIBR artifacts. We can observe them in **Figure 3b**. They are created as the user moves forward (step-in), increases the resolution (zoom), or observes slanted objects. Those cracks correspond to pixels in the target that do not have a preimage in the input view (i.e., no input pixel is mapped to them). However, as their neighboring pixels have a preimage, their color can be interpolated. In other words, the input pixels should be mapped to more than one pixel to compensate for this effect. This can be done using superpixels [20], adapting the pixels size to the camera movement [21], or linking neighboring pixels for rasterization [15, 16] (chosen solution in RVS: adjacent pixels are grouped into triangles that are colorized).

Even if increasing the number of input images can reduce the number of disocclusions, it brings new challenges, as those views need to be consistent in color, in estimated geometry, and in estimated pose. Notably, the depth estimation and the blending of multiple views together rely on consistent colors between the images. As not all camera sensors are equal, small differences in color rapidly generate incoherent depth estimations or nonhomogeneous color patches during view blending (**Figure 3c**). Color correction is usually needed prior to the view synthesis [22, 23] or during the blending step [24, 25].

Moreover, as DIBR relies on the depth information, errors in the depth estimation, a misalignment between the color image and the depth map, or errors in the camera pose estimation lead to ghosting artifacts. When several views are blended together, these artifacts make the objects or their borders appear doubled (**Figure 3d**). A depth map refinement [26, 27] is one way to solve this problem. Another is to choose weighted blending coefficients based on the reliability of each input image [11, 16] (chosen solution in RVS).

Finally, DIBR is structurally limited to the rendering of diffuse objects. Indeed shifting the pixels in the function of their depth assumes that they do not present view-dependent aspects, such as transparency or specularity. When such objects, so-called non-Lambertian, are present in the scene, the linear hypothesis in pixel displacement in the function of the camera displacement is not valid anymore. Adapting the DIBR principles to non-Lambertian objects is nevertheless possible by exploiting additional information, such as structure, normal, and indexes of refraction [28], or a more accurate approximation of the pixel displacement [29–31] (chosen solution in RVS).

2.3 RVS in practice

The DIBR software RVS is designed to render novel viewpoints from any number of input images and depth maps and their camera parameters, without suffering from the above limitations. In order to create a novel view, the input images are warped sequentially. The obtained result is then blended into an image accumulating the outputs of each reprojected input image. This pipeline is shown in **Figure 4**. The warping and blending operations are performed alternatively for each input image using OpenGL [32] or on the CPU [15].

To obtain high-quality results, it is recommended to select candidate input views properly. Therefore, the first step in RVS is an optional view selection. The *n* views the

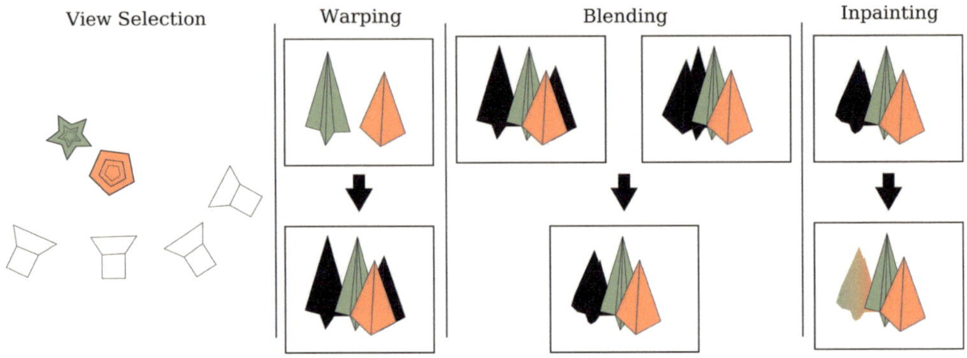

Figure 4.
Overview of the processing pipeline. (1) view selection (optional), (2) warping, (3) blending, (4) Inpainting (optional).

closest to the target image are selected in order to reduce the computation time. Otherwise, all the input images are used to create a new viewpoint.

The second step in RVS is the warping phase. Each input image is divided into a grid of triangles whose vertices are adjacent pixels (**Figure 5a**). Each of these vertices is reprojected to fit to the new camera pose and parameters (**Figure 5b**) and rasterized to avoid the cracks artifacts of **Figure 3b**. Then, each new triangle is given a score that will be used in the blending phase. This score describes the quality of a warped triangle—if the pixels lie on a disocclusion area, their triangle will be stretched and should hence be discarded from the final result (black areas in **Figure 4**-warping and **Figure 5c**). The remaining triangles are then rasterized according to their vertex color in the input image (**Figure 5c**). A depth test prioritizes the pixels with the lowest depth.

When the input images are warped to the target viewpoint, the results need to be blended into one single image. For a given pixel, the final output color c is the weighted mean of the color c_i of each warped input:

$$c = \frac{1}{\sum_i w_i} \sum_i w_i c_i \tag{7}$$

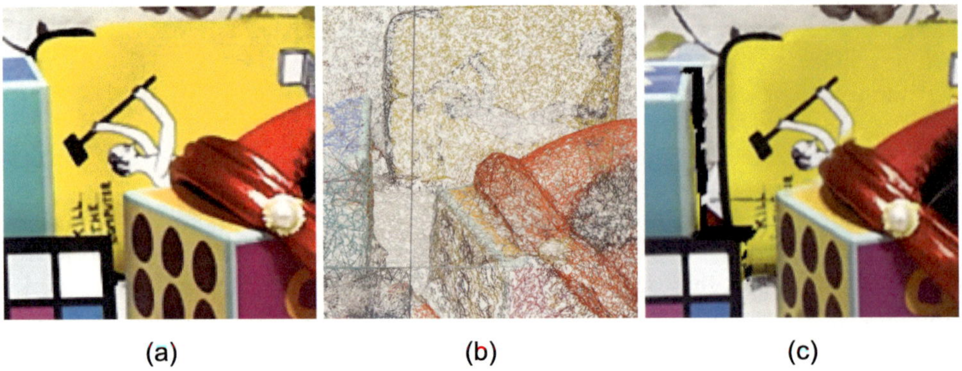

Figure 5.
Adjacent pixels of an input image (a) are grouped into triangles independently of their depth before being reprojected to their new image location (b). Triangles detected as lying on a disocclusion are discarded, resulting in a new warped image (c).

where w_i is a weight representing the quality of a triangle [16], prioritizing foreground objects and highest quality triangles.

Finally, as shown in the inpainting of **Figure 4**, when multiple views are blended together, several occluded regions remain; if the occlusions are small enough, a basic inpainting process can be applied to remove them. Of course, the quality of the inpainting can compromise the overall image quality, hence, inpainting is not recommended. In RVS, the inpainting is not automatic but can be activated. In that case, the empty pixels take the color of the nearest non-empty pixel.

2.3.1 Non-Lambertian case

In the general case, DIBR uses depth maps to predict pixel displacement. However, a point on a non-Lambertian surface does not have a proper color (its color can rapidly change with a change in viewing direction); its appearance is a function of the surrounding scene, the normal at that point, and the index of refraction for refractive surfaces (see **Figure 6**). This not only makes depth estimation through stereo matching impossible but also implies that even with a correct depth map, the object cannot be rendered by a simple pixel shifting.

Alternatively, to model the non-Lambertian surface in itself, it is possible to track its feature movements on the surface [29, 33, 34]. DIBR can be generalized to non-Lambertian objects by replacing the usual depth maps with the coefficients of a polynomial approximating the non-Lambertian features displacement [30, 31]. To clearly understand what this means, let us start with what happens for diffuse objects, where for a lateral camera movement (x, y), the new position (u, v) of a pixel (u_0, v_0) is given by:

$$\begin{pmatrix} u \\ v \end{pmatrix} = \begin{pmatrix} u_0 \\ v_0 \end{pmatrix} + \frac{f}{D}\begin{pmatrix} x \\ y \end{pmatrix}. \tag{8}$$

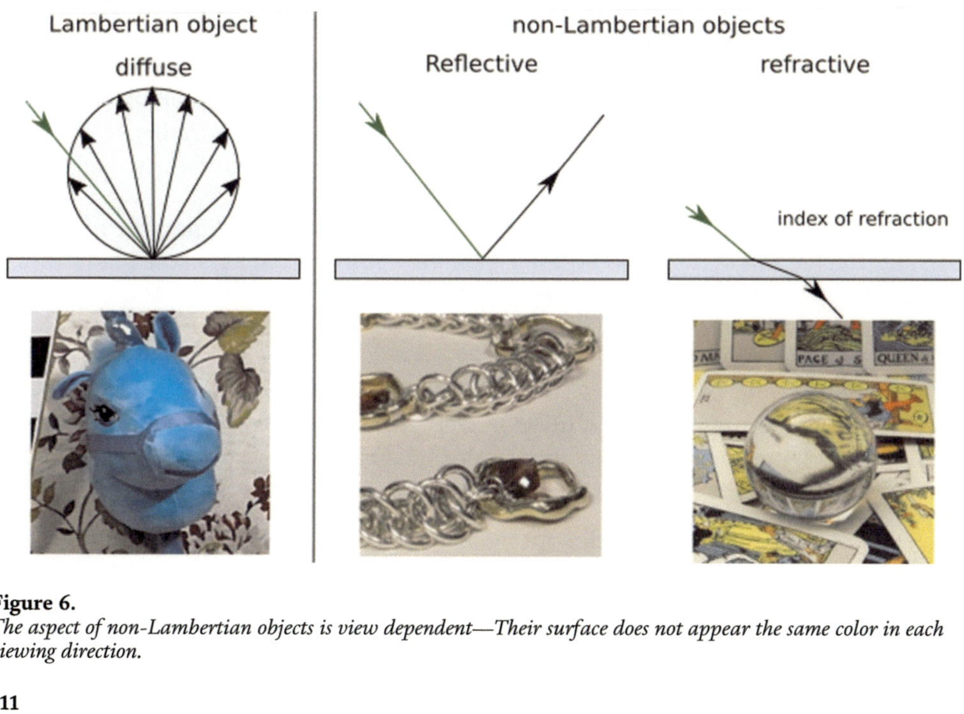

Figure 6.
The aspect of non-Lambertian objects is view dependent—Their surface does not appear the same color in each viewing direction.

We extend this equation for non-Lambertian objects using polynomials:

$$\begin{pmatrix} u \\ v \end{pmatrix} = \begin{pmatrix} u_0 \\ v_0 \end{pmatrix} + \begin{pmatrix} P_u(x,y) \\ P_v(x,y) \end{pmatrix}, \tag{9}$$

with $P_u(x,y) = \sum_i \sum_j a_{ij} x^i y^j$ and $P_v(x,y) = \sum_i \sum_j b_{ij} x^i y^j$. Clearly, the diffuse case corresponds to $a_{1,0} = b_{0,1} = \frac{f}{D}$ and all other coefficients a_{ij}, b_{ij} set to zero.

Consequently, Eq. 9 approximates by a polynomial the nonlinear displacement of a refracted or reflected feature moving on non-Lambertian objects.

However, the polynomial expression rapidly diverges in extrapolation (e.g., when synthesizing a target view that is outside of the input images' hull). The computed feature displacement becomes greater than the inverse of the non-Lambertian object's depth, making the feature to be rendered outside of the non-Lambertian surface. This approximation is hence designed for interpolation and small extrapolation only.

Furthermore, these polynomials are not directly related to the physical reality of the non-Lambertian object. Hence, contrary to the simple relation linking the depth to the disparity of a diffuse object cf. **Figure 2** and Eq. (1), the polynomials of Eq. (9) do not give the object geometry or the index of refraction.

The polynomial is rather designed to "track" non-Lambertian features that move nonlinearly across the input images. It nevertheless encounters the following limitations. For content with semi-transparent objects, the maps should be divided into several layers before applying the polynomial or depth image-based rendering. Scenes with glints and glossiness make it difficult to track features on their surfaces, often leading to a failure case of the proposed method.

3. Reference view synthesis (RVS) software

This section provides practical recommendations for the use of the reference view synthesizer (RVS) [14–16, 32, 35] (https://gitlab.com/mpeg-i-visual/rvs) developed as a DIBR-based view synthesizer for the MPEG immersive video (MIV) standard (https://mpeg-miv.org). Without further details on the compression and storage of immersive content [36], we give a comprehensive method to practically use the software on some test sequences (also provided to the MPEG community while developing RVS).

The following paragraphs give documentation on the image format, the axis system, and the data structure to synthesize new viewpoints from available ready-to-use datasets and/or new content users may provide.

3.1 Input images

RVS can accept any number of input images with depth maps, the only limitation being the computer memory. Each input color image must be provided along with its corresponding depth map.

3.1.1 Color images

The color images can be encoded on three RGB color channels, with 8-bit integers each, in any image format readable by OpenCV, for example, PNG or JPEG format.

Additionally, raw images in YUV can be used, with a bit depth of 8, 10, or 16 bits. In this case, multi-frame raw video can be used, applying the view synthesis on all specified frames.

3.1.2 Depth maps

The depth maps represent the depth coordinate of every point in the image following the forward axis of the camera. Similar to color images, they can be provided in different formats. They have to match the resolution of the input images, but they use only one channel.

The first option is to use the OpenEXR format. In that case, the software reads the depth value in float and uses it directly for reprojection.

In the case of integer coded formats, such as YUV or PNG, the precision can be set to 8, 10, or 16 bit per depth value—the bit depth. YUV files can be encoded in YUV420 or YUV400 format, only the Y channel being used. However, the quantization does not allow to directly use the integer as a depth value. Indeed, it would be impossible to use a depth map in meter units for objects in the range of a few meters or centimeters from the cameras.

To overcome this problem, the depth value is encoded into MPEG's disparity format, mapping the closest object to $2^{\text{bitdepth}} - 1$, and the farthest to 1. To obtain the actual depth value, first we divide the encoded depth map value by $2^{\text{bitdepth}} - 1$ to obtain a value d in the range of $[0, 1]$, then remap the value in the range $[n,f]$ using:

$$d' = \frac{f \times n}{n + d \times (f - n)}. \tag{10}$$

With n and f the near and far values of the scene and d' the depth value lying in $[n,f]$.

For very far objects, this equation is simplified ($f \geq 1000$) to

$$d' = \frac{n}{d} \tag{11}$$

The value 0 in the encoded depth maps is considered as invalid depth. It corresponds, for example, to disocclusions in a depth-sensing device-acquired map.

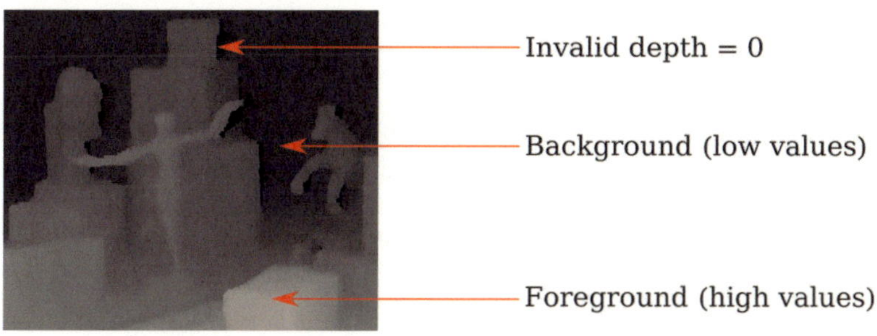

Figure 7.
Encoded depth map on integer values. Due to the shift between the color sensor and the depth sensor, the depth map reprojected to the color image misses some information, leaving invalid pixels, encoded on 0. The foreground objects are encoded on high disparities, while the background objects are encoded on low disparities.

Figure 7 shows an encoded depth map with invalid pixels and objects at different depths. Clearly, the foreground has high values, which corresponds to being a disparity value, that is, the inverse of a depth, cf. Eq. (11).

In the case of polynomial maps for non-Lambertian objects, it is possible to encode up to degree 3 polynomials, hence 18 coefficients, and pass an additional depth map and mask for the non-Lambertian objects. Those coefficients are encoded similarly to the depth maps, using EXR (directly the float value) or YUV (normalized) format. The polynomial maps are numbered from 0 to 19 as follows.

$$\begin{aligned} P_u(x,y) &= a_0x^3 + a_1x^2y + a_2xy^2 + a_3y^3 + a_4x^2 + a_5xy + a_6y^2 + a_7x + a_8y, \\ P_v(x,y) &= b_0x^3 + b_1x^2y + b_2xy^2 + b_3y^3 + b_4x^2 + b_5xy + b_6y^2 + b_7x + b_8y \end{aligned} \tag{12}$$

with a_i corresponding to the map i and b_i to the map $10 + i$. The remaining map 9 is used to encode the depth map for Lambertian objects and the map 19 is used as a mask representing non-Lambertian objects (0 for Lambertian, 1 for non-Lambertian). The coefficients not used are left to 0. If the coefficients are encoded in YUV format, the depth (map n°9) is normalized using Eq. 10, the mask (map n°19) has 0 and 1 values and the other coefficients are linearly normalized between minimal m and maximal M values: $a_{i'} = (M - m)a_i + m$.

3.2 Camera parameters

Additionally to the input images, the camera parameters must be known to create a novel view with DIBR and RVS. The extrinsic parameters describe the position and the rotation of the camera (Eq. 2), while the intrinsic parameters describe the projection matrix (Eq. 3). Perspective and equirectangular projections are also supported, requiring a slightly different description, as explained hereafter.

3.2.1 Extrinsic parameters

Common graphics processing software and APIs, such as Blender [37], COLMAP [38], OpenGL [39], Vulkan [40], specify their own coordinate system, often admitting different axes and directions, and different image coordinates. Transferring data from one application to the other requires then several coordinate transformation steps, which will be summarized here. We use the Omnidirectional Media Format (OMAF) [41] coordinate system of MPEG-I, combined with yaw-pitch-roll angles.

OMAF is the first industry standard for VR. It specifies the coordinate system used in VR applications, the projection and rectangular region-wise packing methods, the metadata storage, encapsulation, signaling, and streaming of omnidirectional data, and finally the media profiles and presentation profiles. For these reasons, it has been adopted in the camera configuration files of RVS.

The OMAF coordinate system is described in **Figure 8**. The axes are defined as follows:

- X: Back-to-front, forward
- Y: Lateral, left

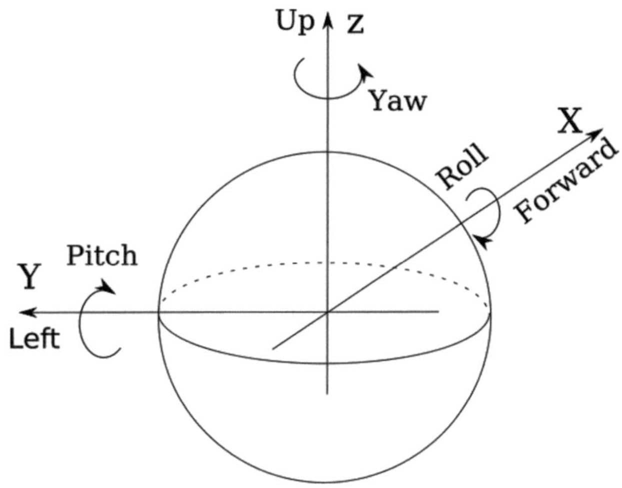

Figure 8.
The omnidirectional media format coordinate system.

- Z: Vertical, up

The rotations in degrees are defined with the Yaw-pitch-roll:

- Yaw: Around the vertical axis
- Pitch: Around the lateral axis
- Roll: Around the back-to-front axis

A camera facing forward has all its rotation angles set to 0. The rotation matrix of the camera (world to camera) in our axis coordinate system is then given by:

$$R = R_z(\text{yaw})R_y(\text{pitch})R_x(\text{roll}) \tag{13}$$

In order to transform a coordinate system from an application to OMAF, one needs to define the coordinate change matrix that matches the three axes, for example:

$$P = \begin{pmatrix} 0 & 0 & -1 \\ 1 & 0 & 0 \\ 0 & -1 & 0 \end{pmatrix} \tag{14}$$

This matrix sets (x', y', z') *(OMAF)* = $(-z, x, -y)$ *(application)*, that is, it represents a coordinate system with the axes (left, down, backward). To transfer from this system to OMAF, we apply it to the rotation and position as follow:

$$\begin{aligned} R' &= P.R.P^T \\ p' &= P.p \end{aligned} \tag{15}$$

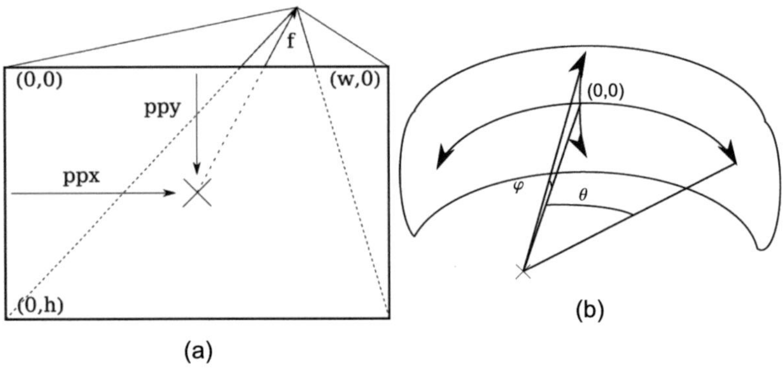

Figure 9.
Intrinsic parameters of the camera for (a) a pinhole projection, (b) an equirectangular projection.

where R' and p' are the rotation and position of the OMAF system, while R and p the rotation and position in the old coordinates.

The unit of the coordinate system does not have any prerequisite but must correspond to one of the depth maps.

RVS handles any number of input and target cameras, each of them can have its own parameters and projection types. In the case of a stereoscopic head-mounted display for VR, two target views – one for each eye – need to be synthesized with a relative position (interpupillary distance) corresponding to the eye distance, usually given by the headset's framework along with the intrinsic parameters.

3.2.2 Intrinsic parameters

The intrinsic parameters can be defined for perspective or equirectangular projections. In both cases, the resolution needs to be specified.

For perspective projection, the input images need to be undistorted. In that way, only the focal length and the principal point need to be specified. Those values are in pixel units, the sensor size corresponding to the image resolution. The focals are given by $\left(f_x, f_y\right)$, corresponding to the horizontal and vertical axis. The principal point $\left(pp_x, pp_y\right)$ is defined from the top-left corner of the image as described in **Figure 9(a)**. A principal point at the center of the image has a value of half the resolution. In the case of equirectangular projection (**Figure 9b**), the horizontal and vertical viewing range must be specified in degrees. For a full 360° panoramic image, the horizontal range is $[-180, 180]$ and the vertical range is $[-90, 90]$. For a 180° image, the horizontal and vertical ranges are $[-90, 90]$.

3.2.3 Camera file

The image specifications and camera parameters are specified in a *json* file with informative headers. An example with a perspective and an equirectangular camera is given here.

Listing 1.1 Cameras calibration file. Cameras.Json.

```
{
  "Version": "3.0",
  "Content_name": "dataset_name",
  "BoundingBox_center": [0, 0, 0],
  "Fps": 30,
  "Frames_number": number_of_frames_in_yuv_file,
  "Informative": {
    [...]
  },
  "lengthsInMeters": true,
  "sourceCameraNames": [
    "input_camera1_name",
    "input_camera2_name",
    [...]
  ],
  "cameras": [
    {
      "Name": "input_camera1_perspective_name",
      "Position": [x, y, z],
      "Rotation": [ yaw, pitch, roll ],
      "Depthmap": 1,
      "Background": 0,
      "Depth_range": [near, far],
      "Resolution": [w, h],
      "Projection": "Perspective",
      "Focal": [ fx, fy],
      "Principle_point": [ ppx, ppy],
      "BitDepthColor": 8,
      "BitDepthDepth": 8,
      "ColorSpace": "YUV420",
      "DepthColorSpace": "YUV420"
    },
    {
      "Name": "input_camera2_equirectangular_name",
      "Position": [x, y, z],
      "Rotation": [ yaw, pitch, roll ],
      "Depth_range": [near, far],
      "Resolution": [w, h],
      "Projection": "Equirectangular",
      "Hor_range": [ θmin, θmax],
      "Ver_range": [ Φmin, Φmax],
      [...]
    },
    {
      [...]
    }
  ]
}
```

An optional parameter, DisplacementMethod, can be set to Polynomial instead of default parameter Depth to specify that, instead of a depth map (Eq. 10), RVS reads a displacement map (Eq. 12). In that case, similarly to the Depth_range, a Multi_depth_range can be specified for the polynomial coefficients in YUV format.

3.3 View synthesis file

In order to perform the view synthesis, an experiment setup file is created. It gives camera views specifications (which views to synthesize given the input viewpoints) in an easy to use *json* format. The file contains:

- Input and target camera parameters file paths—path to the camera file described in the previous subsection. The same file can be used twice if all the input and target cameras are in the same file;
- Input and target camera names matched with the camera names contained in the camera files. Any number of inputs and outputs can be specified;
- Input images, output images, and depth maps file paths. In the case of polynomial maps, numbered from 0 to 19, the number is replaced by a *;
- Number of output frames. Useful for uncompressed YUV video files. The synthesized number of frames can exceed the number of frames in the input videos by specifying an optional NumberOfOutputFrames. In that case, the video will be played back and forth until the desired number of frames is reached;
- Precision: super-resolution factor to reach sub-pixel accuracy;
- Colorspace: internal working color space, can be YUV or RGB. Following the color space used, the result may present small color variations;
- Blending specifications: the method can be Simple (for CPU and GPU usage) or Multispectral (for CPU). Multispectral blending detects the borders in the images, to blend them with a hard threshold and therefore avoids ghosting. The factor represents the power on the weights of Eq. 7.

Listing 1.2. Experiment configuration file. Experiment.Json.

```
{
  "Version": "2.0",
  "InputCameraParameterFile":   "input_cameras.json",
  "VirtualCameraParameterFile": "target_cameras.json",
  "InputCameraNames": [
        "input_camera1_name",
        "input_camera2_name"
  ],
```

```
{
  "Version": "2.0",
  "InputCameraParameterFile":   "input_cameras.json",
  "VirtualCameraParameterFile": "target_cameras.json",
  "InputCameraNames": [
        "input_camera1_name",
        "input_camera2_name"
  ],
  "VirtualCameraNames": [
    "output_camera1_name",
    "output_camera2_name"
  ],
  "ViewImageNames": [
    "path_to_texture1.yuv",
    "path_to_texture2.yuv"
  ],
  "DepthMapNames": [
    "path_to_depthmap1.yuv",
    "path_to_depthmap2.yuv"
  ],
  "OutputFiles": [
    "path_to_output_filename1.yuv",
    "path_to_output_filename2.yuv",
  ],
  "StartFrame": 0,
  "NumberOfFrames": number_of_frames_to_synthesize,
  "Precision": 1.0,
  "ColorSpace": "RGB",
  "BlendingMethod": "Simple",
  "BlendingFactor": 5.0
}
```

3.4 Datasets

To test the view synthesis, we provide references to datasets that are provided with their cameras configuration *json* files. Publicly available datasets are available at the following addresses, while others have been provided as test scenes for MPEG-I immersive video exploration and standardization activities.

- Toystable [42, 43]: a natural dataset with perspective cameras (**Figure 10a**) https://zenodo.org/record/5055542

- Magritte [44–48]: a synthetic dataset with polynomial non-Lambertian maps (**Figure 10b**) https://zenodo.org/record/4488242, https://zenodo.org/record/5047238, https://zenodo.org/record/5047676, https://zenodo.org/record/5047769

- Rabbit [49, 50]: the subaperture views of a multi-plenoptic camera dataset (**Figure 10c**) https://zenodo.org/record/5053770

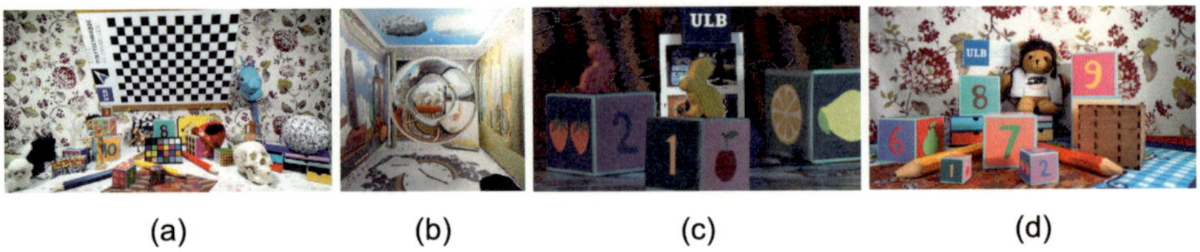

Figure 10.
Overview of the open-source datasets. (a) Toystable consists of two camera arrays at 25 cm (5 × 5 cameras) and 55 cm (3 × 5 cameras) from the scene. (b) Magritte is a 21 × 21 camera array. (c) Rabbit is a 3 × 7 array of 5 × 5 subaperture images of a plenoptic camera. (d) Bear (4 × 8 cameras) is a dataset captured by a Lidar camera with estimated and sensed depth maps.

- Bear [51]: a natural dataset with perspective cameras with estimated and Lidar-sensed depth maps (**Figure 10d**): https://zenodo.org/record/5047463
- MPEG test sequences (including [52–55]): http://mpegx.int-evry.fr/mpegcontent/

4. Displays

We provide in **Figure 11** results obtained with the RVS software on various display types—autostereoscopic or light field screen (Courtesy ETRO-VUB, Belgium), holographic stereograms [56], and head-mounted displays. Additional videos can be found at the following links: https://youtu.be/ikJb9JaaE54 (holographic stereogram) and https://youtu.be/vavw-TcbHf4 (head-mounted-display).

Displaying a dynamic scene in VR requires real-time view synthesis, preferably at 90 frames per second and at a minimum of 30 frames per second for each eye.

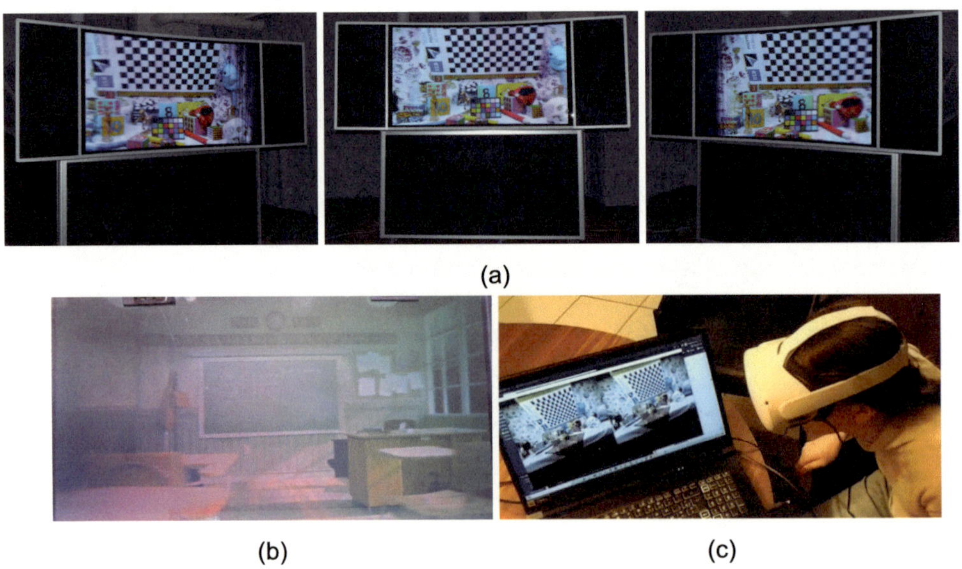

Figure 11.
Instead of acquiring the 100 of views needed for the different kinds of display, RVS recreates them using four input images. (a) Autostereoscopic screen, (b) holographic stereogram, (c) head-mounted display.

		Number of input images							
Dataset	**Resolution**	**1**	**2**	**3**	**4**	**5**	**6**	**7**	**8**
Toystable [42, 43]	1920 × 1080	90	51	46	49	37	36	31	29
Fencing [52]	1920 × 1080	90	90	88	56	52	47	46	40
Painter [53]	2048 × 1088	90	90	90	80	60	56	53	46
Museum [54]	2048 × 2048	52	53	47	31	22	21	16	16
Classroom [55]	4096 × 2048	55	31	30	19	15	11	9	8

Table 1.
The frame rate for view synthesis in VR depends on the number of input images and their resolution. The output images all have the resolution of oculus rift (i.e., 1080×1200 pixels). Those results have been obtained on a windows PC with Intel Xeon E5–2680@2.7GHz CPU and NVIDIA GTX 1080TI GPU.

However, the processing time depends on the number of input images and their resolution – since their pixels form the mesh – resulting in different frame rates [16] (see **Table 1**). Using a NVIDIA GTX 1080TI GPU, around four input images at a full HD resolution can be processed to obtain a high visual quality while reaching real-time navigation.

In the case of the currently developed light field head-mounted display [57], the constraint is double—in addition to the real-time requirement, all the light rays reaching the user's pupils need to be displayed to make the eye accommodation possible on the close objects, that is, not only one image per eye but all the micro-parallax views around the eye position are rendered.

4.1 Additional tools

In this section, we provide references for additional tools, which are not directly involved in the view synthesis but are nevertheless useful to prepare a dataset.

4.1.1 Camera calibration

The first step prior to DIBR is finding the camera parameters. Accurate intrinsic parameters, including distortion parameters, can be found using a calibration checkerboard-pattern, if the scene has a large enough baseline, or directly during the scene reconstruction (structure-from-motion (SfM) with the retrieval of intrinsic parameters). Using a pattern gives more accurate results but requires a supplementary preprocessing step. There exists open-source software such as Kalibr [58] and OpenCV [59] for camera calibration.

Extrinsic parameters of a set of cameras are retrieved using SfM, with or without the intrinsic parameters known as prior [60]. There exist many open-source software such as COLMAP [38] or AliceVision [61]. Those softwares calibrate the camera and automatically undistort the images.

4.1.2 Depth estimation

Besides parameters estimation, DIBR requires corresponding depth maps for each input view. If they are not acquired with a depth-sensing device, they can be computed using stereo-matching algorithms. Among many algorithms, Depth Estimation

Reference Software (DERS) [62] and Immersive Video Depth Estimation (IVDE) [63, 64] are recognized by the MPEG-I community.

5. Conclusions

In this chapter, an overview of the main steps and frequent problems of view synthesis are described. By starting from sparse input pictures, we showed a DIBR method that renders the parallax effect on a multitude of displays, allowing a user to experience new aspects of multimedia immersion. In the second part, a description of how one can start experimenting with the state-of-the-art RVS software is thoroughly explained to avoid common pitfalls.

As research progresses, novel methods to create view synthesis emerge, such as NeRF, however, recent research results demonstrate that DIBR methods will still reach high-quality performances [16], in real time, that will be highly applicable in immersive applications, for example, in the context of MPEG immersive video.

Acknowledgements

This work was supported in part by the Fonds de la Recherche Scientifique – FNRS, Belgium, under Grant n·33679514, ColibriH; and in part by the HoviTron project that received funding from the European Union's Horizon 2020 research and innovation program under grant agreement N·951989 (https://www.hovitron.eu/).

Conflict of interest

The authors declare no conflict of interest.

Abbreviations

CPU	Central Processing Unit
DERS	Depth Estimation Reference Software
DIBR	Depth image-based rendering
GPU	Graphics Processing Unit
IVDE	Immersive Video Depth Estimation
MIV	MPEG Immersive Video
MPEG-I	Moving Picture Experts Group-Immersive
NeRF	Neural Radiance Field
OMAF	Omnidirectional Media Format
RGB	Red-Green-Blue format
RVS	Reference View Synthesizer
SfM	Structure-from-Motion
VR	Virtual Reality
YUV	Luminance-Chrominance format

Author details

Sarah Fachada†, Daniele Bonatto†, Mehrdad Teratani and Gauthier Lafruit*
Université Libre de Bruxelles, Brussels, Belgium

*Address all correspondence to: gauthier.lafruit@ulb.be

† These authors contributed equally.

References

[1] Euclid of Alexandria. Optics; 300 BC.

[2] Al-Haytham HI. Book of Optics. Vol. I-VII circa 1027

[3] Renner E. Pinhole Photography: From Historic Technique to Digital Application. 4th ed. Amsterdam, Boston: Focal Press; 2009

[4] Maeda T, Suenaga R, Suzuki K, Panahpour Tehrani M, Takahashi K, Fujii T. Free Viewpoint Video for Sports Events Using Multi-Resolution Visual Hull and Micro-Facet Billboarding. Proc. Intl Workshop on Smart Info-media System in Asia, Ayutthaya, Thailand; 2016

[5] Suenaga R, Suzuki K, Tezuka TP, Tehrani M, Takahashi K, Fujii T. A practical implementation of free viewpoint video system for soccer games. In: Three-Dimensional Image Processing, Measurement (3DIPM), and Applications. Vol. 9393. International Society for Optics and Photonics, San Francisco, CA, USA; 2015, 2015. p. 93930G

[6] Hartley R, Zisserman A. Multiple view geometry in computer vision. 2nd ed. Cambridge, UK, New York: Cambridge University Press; 2004

[7] Agarwal S, Furukawa Y, Snavely N, Simon I, Curless B, Seitz SM, et al. Building Rome in a Day. New York, USA: Communications of the ACM. 2011;**54**(10):105

[8] Mildenhall B, Srinivasan PP, Tancik M, Barron JT, Ramamoorthi R, Ng R. NeRF: Representing Scenes as Neural Radiance Fields for View Synthesis. In: Vedaldi A, Bischof H, Brox T, Frahm JM, editors. Computer Vision–ECCV 2020. Cham: Springer International Publishing; 2020. pp. 405-421

[9] Fehn C. Depth-image-based rendering (DIBR), compression, and transmission for a new approach on 3D-TV. In: Stereoscopic Displays and Virtual Reality Systems XI. Vol. 5291. International Society for Optics and Photonics, San Jose, CA, USA; 2004. pp. 93-105

[10] Penner E, Zhang L. Soft 3D reconstruction for view synthesis. ACM Transactions on Graphics. 2017;**36**(6): 1-11

[11] Mildenhall B, Srinivasan PP, Ortiz-Cayon R, Kalantari NK, Ramamoorthi R, Ng R, et al. Local Light Field Fusion: Practical View Synthesis with Prescriptive Sampling Guidelines. ACM Transactions on Graphics. 2019;**38**(4). Art. 29, pp. 1-14

[12] Vagharshakyan S, Bregovic R, Gotchev A. Light field reconstruction using shearlet transform. IEEE Transactions on Pattern Analysis and Machine Intelligence. 2018;**40**(1): 133-147

[13] Wu G, Zhao M, Wang L, Dai Q, Chai T, Liu Y. Light field reconstruction using deep convolutional network on EPI. In: 2017 IEEE Conference on Computer Vision and Pattern Recognition (CVPR). Honolulu, HI: IEEE; 2017. pp. 1638-1646

[14] Kroon B. Reference View Synthesizer (RVS) Manual [N18068]. ISO/IEC JTC1/SC29/WG11, Ljubljana, Slovenia; 2018

[15] Fachada S, Bonatto D, Schenkel A, Lafruit G. Depth Image Based View Synthesis With Multiple Reference Views For Virtual Reality. In: IEEE 2018-3DTV-Conference: The True Vision-Capture, Transmission and

Display of 3D Video (3DTV-CON). Helsinki: IEEE; 2018

[16] Bonatto D, Fachada S, Rogge S, Munteanu A, Lafruit G. Real-Time Depth Video Based Rendering for 6-DoF HMD Navigation and Light Field Displays. IEEE Access. 2021;**9**: 146868-146887

[17] Li S, Zhu C, Sun MT. Hole filling with multiple reference views in DIBR view synthesis. IEEE Transactions on Multimedia. 2018;**20**(8):1948-1959

[18] Telea A. An image inpainting technique based on the fast marching method. Journal of Graphics Tools. 2004;**9**(1):23-34

[19] Huang HY, Huang SY. Fast hole filling for view synthesis in free viewpoint video. Electronics. 2020;**9**(6): 906

[20] Tehrani PM, Tezuka T, Suzuki K, Takahashi K, Fujii T. Free-viewpoint image synthesis using superpixel segmentation. APSIPA Transactions on Signal and Information Processing. 2017; **6**, e5, pp. 1-12

[21] Senoh T, Tetsutani N, Yasuda H. Depth Estimation and View Synthesis for Immersive Media. 2018 International Conference on 3D Immersion (IC3D), Brussels, Belgium; 2018. pp. 1-8

[22] Reinhard E, Adhikhmin M, Gooch B, Shirley P. Color transfer between images. IEEE Computer graphics and applications. 2001;**21**(5):34-41

[23] Fecker U, Barkowsky M, Kaup A. Histogram-based prefiltering for luminance and chrominance compensation of multiview video. IEEE Transactions on Circuits and Systems for Video Technology. 2008; **18**(9):1258-1267

[24] Dziembowski A, Domański M. Adaptive Color Correction In Virtual View Synthesis. Stockholm: 2018-3DTV-Conference: The True Vision-Capture, Transmission and Display of 3D Video (3DTV-CON), IEEE; 2018

[25] Dziembowski A, Mieloch D, Różek S, Domański M. Color Correction for Immersive Video Applications. IEEE Access. 2021;**9**:75626-75640

[26] Vijayanagar KR, Loghman M, Kim J. Refinement of depth maps generated by low-cost depth sensors. In: 2012 International SoC Design Conference (ISOCC), Jeju, Korea, IEEE; 2012. p. 355-358

[27] Sancho J, Sutradhar P, Rosa G, Chavarrías M, Perez-Nuñez A, Salvador R, et al. GoRG: Towards a GPU-Accelerated Multiview Hyperspectral Depth Estimation Tool for Medical Applications. Sensors. 2021;**21**(12):4091

[28] Lochmann G, Reinert B, Ritschel T, Müller S, Seidel HP. Real-time Reflective and Refractive Novel-view Synthesis. 19th International Workshop on Vision, Modeling and Visualization (VMV), Darmstadt, Germany. Eurographics Association, 2014. pp. 9-16.

[29] Nieto G, Devernay F, Crowley J. Linearizing the Plenoptic Space. In: 2017 IEEE Conference on Computer Vision and Pattern Recognition Workshops (CVPRW). Honolulu, HI, USA: IEEE; 2017. pp. 1714-1725

[30] Fachada S, Bonatto D, Teratani M, Lafruit G. Light field rendering for non-Lambertian objects. Electronic Imaging. Stereoscopic Displays and Applications XXXII. 2021, pp. 54-1-54-8

[31] Fachada S, Bonatto D, Teratani M, Lafruit G. Polynomial Image-Based Rendering for non-Lambertian Objects.

Munich, Germany: Visual Communication and Image Processing 2021; 2021. p. 5

[32] Bonatto D, Fachada S, Lafruit G. RaViS: Real-time accelerated View Synthesizer for immersive video 6DoF VR. Burlingame, USA: Society for Imaging Science and Technology (IS&T) - Electronic Imaging; 2020

[33] Maeno K, Nagahara H, Shimada A, Taniguchi RI. Light Field Distortion Feature for Transparent Object Recognition. In: 2013 IEEE Conference on Computer Vision and Pattern Recognition. Portland, OR, USA: IEEE; 2013. pp. 2786-2793

[34] Xu Y, Nagahara H, Shimada A, Ri T. Transcut: Transparent object segmentation from a light-field image. Proceedings of the IEEE International Conference on Computer Vision, Santiago, Chile; 2015. pp. 3442-3450

[35] Bonatto D, Fachada S, Lafruit G. RVS kickstart: example test sequences and config files. Brussels, Belgium: ULB DI-fusion; 2019

[36] Lafruit G, Bonatto D, Tulvan C, Preda M, Yu L. Understanding MPEG-I Coding Standardization in Immersive VR/AR Applications. SMPTE Motion Imaging Journal. 2019;**128**(10): 33-39

[37] Blender Online Community. Blender - A 3D Modelling and Rendering Package. Blender Institute, Amsterdam: Blender Foundation; 2021 Available from: http://www.blender.org

[38] Schonberger JL, Frahm JM. Structure-from-Motion Revisited. In: 2016 IEEE Conference on Computer Vision and Pattern Recognition (CVPR). Las Vegas, NV, USA: IEEE; 2016. pp. 4104-4113

[39] Kessenich J, Baldwin D, Rost R. The OpenGL® Shading Language, Version 4.60.7, The Khronos Group Inc.; 2019

[40] Group TKVW. Vulkan® 1.2.202 - A Specification; 2021.12.07.

[41] Omnidirectionnal Media Format. MPEG-I; 2021 Available from: https://mpeg.chiariglione.org/standards/mpeg-i/omnidirectional-media-format

[42] Bonatto D, Schenkel A, Lenertz T, Li Y, Lafruit G. ULB High Density 2D/3D Camera Array data set, version 2 [M41083]. ISO/IEC JTC1/SC29/WG11; 2017

[43] Bonatto D, Fachada S, Lafruit G. ULB ToysTable. Zenodo; 2021. Available from: https://zenodo.org/record/5055543.

[44] Fachada S, Bonatto XY, Teratani M, Lafruit G. Two Non-Lambertian Test Materials and View Synthesis Performance [m56450]. ISO/IEC JTC1/SC29/WG11; 2021

[45] Fachada S, Bonatto D, Teratani M, Lafruit G. Transparent Magritte Test Sequence. Zenodo; 2021 Available from: https://zenodo.org/record/4488243

[46] Fachada S, Bonatto D, Teratani M, Lafruit G. Mirror Magritte Torus Test Sequence; 2021. Available from: https://zenodo.org/record/5048262

[47] Fachada S, Bonatto D, Teratani M, Lafruit G. Magritte Sphere Test Sequence; 2021. Available from: https://zenodo.org/record/5048265

[48] Fachada S, Bonatto D, Teratani M, Lafruit G. Magritte Sphere Video Test Sequence; 2021. Available from: https://zenodo.org/record/5048270

[49] Fachada S, Xie Y, Bonatto D, Lafruit G, Teratani M. [DLF] Plenoptic 2.0 Multiview Lenslet Dataset and Preliminary Experiments [m56429]. ISO/IEC JTC1/SC29/WG11; 2021

[50] Fachada S, Yupeng X, Bonatto D, Lafruit G, Teratani M. RabbitStamp Test Sequence; 2021. Available from: https://zenodo.org/record/5053770

[51] Xie Y, Fachada S, Bonatto D, Lafruit G. HoviTronBear Test Sequence; 2021. Available from: https://zenodo.org/record/5047464

[52] Domański M, Dziembowski A, Grzelka A, Mieloch D, Stankiewicz O, Wegner K. Multiview test video sequences for free navigation exploration obtained using pairs of cameras [M38247]. ISO/IEC JTC1/SC29/WG11; 2016

[53] Doyen D, Langlois T, Vandame B, Babon F, Boisson G, Sabater N, et al. Light field content from 16-camera rig [M40010]. ISO/IEC JTC1/SC29/WG11; 2017

[54] Doré R, Briand G, Tapie T. Technicolor 3DoFPlus Test Materials [M42349]. ISO/IEC JTC1/SC29/WG11; 2018

[55] Kroon B. 3DoF+ test sequence ClassroomVideo [M42415]. ISO/IEC JTC1/SC29/WG11; 2018

[56] Fachada S, Bonatto D, Lafruit G. High-quality holographic stereogram generation using four RGBD images. Applied Optics. 2021;60(4):A250–A259. Publisher: Optical Society of America.

[57] Bonatto D, Hirt G, Kvasov A, Fachada S, Lafruit G. MPEG Immersive Video tools for Light-Field Head Mounted Displays. Munich, Germany: IEEE International Conference on Visual Communications and Image Processing; 2021. p. 2

[58] Furgale P, Rehder J, Siegwart R. Unified temporal and spatial calibration for multi-sensor systems. In: 2013 IEEE/RSJ International Conference on Intelligent Robots and Systems. Tokyo: IEEE; 2013. pp. 1280-1286

[59] Bradski G. The open CV library. Dr Dobb's Journal: Software Tools for the Professional Programmer. 2000;25(11): 120–123;Publisher: Miller Freeman Inc.

[60] Quan L, Lan Z. Linear N-point camera pose determination. IEEE Transactions on Pattern Analysis and Machine Intelligence. 1999;**21**(8): 774-780

[61] Griwodz C, Gasparini S, Calvet L, Gurdjos P, Castan F, Maujean B, et al. AliceVision Meshroom: An open-source 3D reconstruction pipeline. Proceedings of the 12th ACM Multimedia Systems Conference - MMSys '21. ACM Press; 2021

[62] Rogge S, Bonatto D, Sancho J, Salvador R, Juarez E, Munteanu A, et al. MPEG-I Depth Estimation Reference Software. In: 2019 International Conference on 3D Immersion (IC3D). Brussels, Belgium: IEEE; 2019. pp. 1-6

[63] Mieloch D, Stankiewicz O, Domański M. Depth Map Estimation for Free-Viewpoint Television and Virtual Navigation. IEEE Access, Conference Name: IEEE Access. 2020;**8**:5760-5776

[64] Mieloch D, Dziembowski A. Proposal of IVDE 3.0 [m55751]. ISO/IEC JTC1/SC29/WG11; 2020

Chapter 7

Leveraging on Data Sciences: Review of Architectural Practice and Education in Nigeria

John Allison and Anita Alaere Bala

Abstract

Big data sciences demand the significant role of the architect. Particularly, facilitate the birth of an antifragile construction industry and more robust data sciences community of professionals. Skilled community necessary to build sustainable liveable cities with emerging creator's economy. Liveability, well-being, and sense of belonging in the city are connected. Conversely, dismissive attitude by decision-makers towards architectural practice and education, even among architects, in recognizing architecture as data-driven and source of data deserve rethink. Here the chapter demonstrates architects as data scientists and the symbiotic relationship that exist between architecture and 3D computer graphics while highlighting emerging data sciences opportunities and threats. The chapter adopted principally reviews of scholarly literatures, draws from authors' 20-years personal experiences, and industry leaders' views. The language is accessible yet academically concise. The chapter concluded with recommendations, including highlights of big data technologies potential transformation of 3D computer graphics. The implications are policy, design, and education.

Keywords: 3-D computer, graphics, architecture, data science, big data, city, Web 3.0, creators economy

1. Introduction

Do Big Data sciences require any role of the architect? Is there any merit for the dismissive attitude against architectural practice as big data knowledge based? Are the architectural institutions agile and responsive to the transformation opportunities big data sciences are bringing to the architectural education and architects? Is architecture data driven and source of data, including does graphics play any role? The chapter sets out to find the answers. In [1] computer 3D graphics, a term first coined by William Fetter in 1961 to describe his work at Boeing, have played a significant role in the design, engineering, and gaming industry. Nonetheless, the impact on 3D computer graphics, like with all architectural endeavors, of the explosion of big data fuelled by data sciences since 2000 will just be as transformative like the internet (see [2]). While earliest 3D computer graphics can be traced to late 1970s with the 3-D Art Graphics by Kazumasa Mitazawa released in June 1978 is one the earliest [3].

Graphic writing as structured visual expression is not strange in Africa as recognized in [4] "with multiple varied communication tools, including written symbols, religious objects, oral tradition, and body language" deployed in organizing "daily life, enable interactions between human and the natural and spiritual worlds, and preserve and transmit cosmological and cosmological belief systems". Conversely, Big data is more than communication, powered by the data sciences and associated technologies. For instance, in understanding an individual (for medical or housing needs), big data will include all unstructured and structured forms [4] in addition to the individual's medical history and other demographic details. To understand industry for operational optimization, as noted in [5] big data "is turning mobility, medical, automobile, and pharmaceutical billions of gigantic data into actionable insights to help make informed business decisions, including all significant industries. Big Data is providing every person 320 times more information provided by the third century store house of knowledge, the Library of Alexandria [2] Blockchain is providing the shared, secured holding and transparency of big data in never before accelerated ways as in video-1 [6]. Nevertheless, a "multifaceted team with complementary skills are needed to realize the full value of big data" [7].

Conversely, the construction industry seems indifferent in harnessing its big data. A few reasons are:

- Lack of effective leadership [8].

- The lack of understanding of the impact and purpose of data sciences in broader society [9].

- Data sciences perceived narrowness, and quick testing and scrutinizing rather than engaging and inspiring has been a discouraging factor among women. Uber, the public cab transport transformer, and the construction Building Information Modeling (BIM) entered the market about the same period in 2009. Conversely, BIM regardless of the potentials beyond design and 3D modeling as buttressed in [10], it has not impacted the construction industry as much as Uber has done to transportation.

Many things have changed, but the architects' fingers, eyes, and creative intellect are yet valuable means to ends in architecture. The sense organs capture, process, analyse, memorise as part of "database of the mind." The same is share and enjoy or rejected as pattern, images, sounds, and alike. Depending on the creative database, architects can deliver static non-interactive alphanumeric multisensory models before the emergence of computer 3-D graphics and its rapid advancements. Conversely, the imagination, hands, and simple tools like pencil pen, rule, Set Square, and paper, for another 30 years ahead (as in [11]), the architect, will continue express and communicates aided by standard architectural graphic symbols and elements with understanding of design principles. Discussing graphics, Obina Wodike, an architect and Graphic lecturer with Port Harcourt Polytechnic, noted graphics is not new to human civilization. Wodike buttressed his point tracing graphics to the Biblical times, Egyptian Heliographic, and cited Thomson (1993). Thomson connected graphics and the era when historians, sociologists, linguists, and visual communication students produced art for commercial purposes, including mass consumption or graphic design, with 1986 marking the formation of Graphic Design Education. The importance of "origin" as a continuous influencer of the future progress validating the continuous influence of the seemingly primitive tools -architects finger, pencil, practices, and more- was further buttressed in [12]. Stating, "the origin and practice in unmasking assumptions within current forms and practice" [12] while in [13] described graphic navigation to imply the manipulation of graphic devices, including organization, hierarchy, line weights, colors, and more on a "surface" which range from a book to a highly complex computer screen. Organizations must embrace learning culture and there must be effective strategies to leverage on emerging opportunities.

2. Literature review

2.1 Brief description

- Graphics: The Cambridge Dictionary gave for Graphic to mean "very clear and powerful" and "having to do with drawings and images" [14]. It is communication by means of standard elements and symbols with creative understanding and design philosophy [13].
- Data: information, especially facts or numbers, collected to be examined and considered and used to help decision-making, or information in an electronic form that can be stored and used by computer (ibid) [13].
- Data Sciences: "interdisciplinary field that uses scientific methods, processes, algorithm and systems to extract knowledge and insights from noisy, structured, and unstructured data, and apply the knowledge and actionable insights from data across a broad range of application domains" [15] data science is different from information science and communication. As concept the term include "statistics, data analysis, informatics, and their related methods in order to understand and analyses actual phenomena with data" [15].
- Data scientist one who create programming codes and combines it with statistical knowledge to create insights from data [15].
- Small Data: For centuries and up till now, small data has been and is useful. We recall experiment, production quality control, social sciences all using small sample sizes to represent the larger population in their investigations [16]. The data sample size can range from 30 to a value determined by Fisher formula [17]. Hardly over a million, relative to big data with sample size in billions and increasing. IBM Watson went through millions of file samples of a patient within few minutes to determine the unique Leukemia that have baffled doctors using relatively negligible data sample over days [18].
- Big Data: Extremely large data set that that may be analyzed computationally and impossible with traditional method to reveal patters trends and association, especially relating to human behavior and interaction; supported in [19] by products, usually useless and unstructured, but insightful and useful when cleaned, labeled, structured, and stored appropriately for secure and safe retrievals.
- Creators economy: An Web 3.0 which will be in sharp contrast with Web 2.0 and web 1.0. In web 3.0 the content creators are owners of what they create (see [20]) Totally different from the current work place and web 2.0 where one works for JP Morgan, Shell, Exxon, Google, Facebook to help create the mega profits, but have no strategic say in the affairs of the firms. They are paid salary- minute of profit- or noting for their creation; "content "created as it is with Google and

Facebook. One man or few decide what the majority do; what to produce, where to, how to, and who gets what, and how the profit is shared.

- Data Scientist From SAS, a “leader in business analytics software and services” has been a leading global business analytic and intelligence software developer since 1976. The 2021 personality survey of data scientists report in [21] identified several personalities of the data scientist. Four relevant to the chapter are: “Deliverers,” “Drivers,” “Voices,” and “Ground Breakers.” In another publication, in Towards Data Science website, Andrew Lombari (2022) discussed “7 Data Science Projects You Should do to make your resume stand out” (see [22]): Regression project, Classification project, Clustering project, Sentiment analysis project, Recommender system project, Natural Language Processing (NLP) project, Artificial neural network project. The various traits in [21, 22] of data scientist fit the architect discussed elsewhere in [9].

2.2 Transformations and big data

Like hurricane, big data inspired technologies transformations will ultimately blow across the construction industry and only the prepared, agile, and adaptive will survive it. For example:

- IBM Watson machine identified a rear form of Leukemia, including treatments after analyzing millions of data of 60-year patient; that baffled doctors [18].

- In the UK, a Nigerian developed algorithm that identify glaucoma more accurately and swiftly than doctors [23].

- In India, the pharmaceutical industry is joining China to use AI to reduce cost, improve quality and minimize waste [24].

- In Natural Language Processing (NLP) have near perfect human models like WuDao 2.0 and GPT-3 [25].

- In [26] GE Additive Manufacturing (AM) is revolutionizing manufacturing or “mass production of components.” The hand tool-free AM process depends directly on CAD data. It is also called “indirect manufacturing”. MIT Additive Printing, also called 3D Printing, is a process of building an object one thin layer at a time, but with each 3D print head delivering as thermoplastic material deposit multiple layer of the build material. The order of 12000 engines with 19 fuel nozzles per engine for Boeing 737 and the Air Bus 320neo will be achieved by 3D printing (see [26]).

- In animation and simulation, “the divide between the real world and the digital world is getting thinner as shown in **video** [27].

Nevertheless, much of these shall be changed with Quantum computing replacing traditional super computer, as quantum teleporting, and virtual world (virtual reality- VR, virtual augmentation -VA) is fully developed (see [28, 29]). Supporting the view in video-3, Professor Michio Kaku, a physicist discussed the era of “brainet” when “computer chip will cost as little as penny”. A concept beyond fiction, the “brainet” project will mark the era when, like the internet, human brains will be sending signals directly and networking with other brains [30]. In [31] hint the revolutionary changes Johnson (1984) said will occur in computer-aided architectural design process. An instance, "architects will have to learn to think in the computer's terms." Also, in [31] reported uploading or downloading from the human memory is not theoretical but practicable.

3. Architects and architecture

3.1 The architect

The architect is well trained to design buildings and processes of achieving them (see engineer's view of the architect in [32]). Architecture ranks among topmost toughest courses in the world with students averaging 22.2 hours of study time weekly; 5.28 hours sleep per night compared to seven hours expert recommended [33]. Often called "master builder," a phrase traced to the original meaning of the word in Greek and Roman renditions. In "English speaking world" it was traced to 1563 and in 1898 (see [34]) A significant product of the architect is architecture, including landscaping which seen in [35] to include building transformed into a living art and expression; not a style. Architecture defines the limit of the city while in [36] "urban planners influence the entire communities with decisions about how areas are developed and even where building can be placed."

Additionally, leadership in of the building and construction industry may have been ascribed to the architect, but with new solutions from digital construction, computing and simulation, it is being challenged. Indeed, entry of more players from non-traditional construction cores, is therefore challenging the architect's leadership. Greater burden is on the architect and related institutions if the cap must fit. New thoughts in [37] are demanding a sectional leadership such as architecture management, construction management, engineering management. Also, with concept like integrated project Development (IPD) and "profit distribution of IPD" based on "degree of participation." Axiomatic to state, "a finished building represents the abilities of more than one individual" [37].

In another study, evidence supports a significant level of ignorance and indifference in the construction industry (see Appendix 1) [9]. All the same, the study demonstrated architect's leadership among peers in the construction industry with a significant lead of 80 percent in participation. This is despite the equal treatment across selected clusters- architecture, engineering, management, and others [9]. Additionally, the National Building Code of Nigeria, Neufert Architects' Data and 3D [38], ArchiCAD, AutoCAD. Apple Pencil(with iPad Pro) recently ranked among world-top 25 inventions [39], and in [40] Germany's most beautiful hydroelectric plant, all have significant big data input of the architects. All these point to the intersection of architecture and data science. Progressing, the chapter will develop and cover the implications of poor data for cities and city users after critical reviews of over fifty purposed sampled scholarly and appropriate articles. Personal experiences and opinions of industry leaders shall be helpful.

3.2 Ignorance and dismissive attitude

The figures below show the survey carried out in 2021 that gave graphic portraits of both ignorance and dismissive attitude of the architect as a professional creating data and significantly with data sciences skills [9]. See appendix 1.

3.3 3D computers graphics and architecture symbiotic relationship

The design process is initially done with a pencil, a ruler, a scale rule, and paper. Then, the process takes a longer time, and relatively inaccurate and significantly limited. Yet they are marvelous works of the likes of Frank Floyd Wright, and alike. Conversely, limitations were significantly broken with new construction materials, and computer aided design (CAD) gave birth to a greater freedom evidenced in architectural expression clustered under Deconstructivism [41, 42]. Libeskind, Zaha Hadid are few Deconstructivists, see **Figure 1**. The use of modern technologies extends throughout the entire design with distorted surfaces, odd leanings, and fractured wall surfaces. Indeed, possible due to speed, fluidity, and accuracy delivered by CAD which significantly relieved architects of time required for complex forms; for more quality family life or creative endeavors. Also, architectural studios are looking very different compared to a decade ago [42]. In Nigeria, the studios are sluggishly adapting, see **Figure 2**.

Indeed, things have changed with the introduction of CAD and significant architectural practice uses CAD: few are ArchiCAD, Revit, BIM, Chief Architect,

Figure 1.
A fit for Deconstructivism by Zaha Hayed. Credit UNSPLASH.

Figure 2.
Architecture studio: sleepless nights. Credit: Allison.

Lumen360, and Sketchup. These tools make easy the organization of data to be more meaningful visually. Thus, in use more commonly among young people who are quick to practice even as students, heavily patronized by clients seeking for cheap labor and thus, compounding the problem of quackery in Nigeria.

More, 5G, 7D are merging to further reshape the design terrains and how architects work. In [43] the symbiotic partnership (man-computer) will enable "men will set the goals, formulate the hypothesis, determine the criteria, and perform the evaluations. Computing machines will do the routinizable work that must be done to prepare the way for insights and decisions in technical and scientific thinking." Design process is not linear, but cyclic for user experience designer because efforts are continually made satisfy the users leaving assumption behind and uncover alternative solutions" [44]. The CAD tools have made the iterative processes easy because design for real people not machines and robots imply understanding complex processes [41, 44]. Further, it empowers the client to make more informed contribution to the design process. From the lens of digital humanities and national development, 3D Computer Graphics is, for example, contributed to the restored sense of pride and patriotism among Chinese young people [45]. The regeneration of the Summer Palace from ruin has turned the once idle ruins into tourist site. With 3D computer graphics the beauty and splendor of the Summer Palace has been fully animated. The ruin site thronged with Chinese young people restored, reliving the past, appreciating the true stories of Summer Palace [45].

All above, 3D computer graphic is important, and noteworthy is the important role played by architects in identifying standardizing, creating symbols, and elements for extracted from ethnographic studies by architects in use by 3D computer graphic community. The standard elements, symbols, minimum dimensions, and other research driven data stored and published in the 680-page Neufert Architects Data authored by an architect, Ernst Neufert indeed testify to interdependent and mutually benefit from each other's progress (see [38]). More benefits are on the horizon, with Metaverse and creator economy is one the fastest growing sector; in terms of income generation, job creation, and export earnings [20]. View supported in [46] as Internet of value ((IoV), Web3.0, discussed the "risky necessity" "datafication," "dematerialization," "platformisation," in addition to social and resource value awareness" (see [44, 46]) architects as potentially key benefactor will have increasing role to play.

3.4 Architect, as data scientist

Architectural production includes: cities, homes, furniture, and more. In addition, it takes a minimum range of 5-6 year to be qualified to pursue full professional or chartered membership in Nigeria. Thus, architecture is a rigorous and highly educated profession [33] and in [47] architecture as toughest major in America from 2017 survey of global students' survey by students. An architectural study is one with high incidence of sleep deprivation among students, **Figure 2**. While architects rely on quantitative and qualitative analysis of data to recommend and make decisions, architectures consume about 55% global electricity and contribute about 38% of total global energy related Carbon IV Oxide (CO_2) waste into the environment [48]. As by-product of architecture, big data is elucidated below focusing on waste paper generated in office activities and how big data can be captured for recycling.

The consulting office is involved in design through tender processes to construction, including post-construction activities. The activities are usually accompanied the shredded waste papers which remain useless not until the following action is carried out:

- The waste papers associated with the project are weighed;
- Findings (associated data) standardized, labeled (like design, stage, tender stage, construction stage)
- Related to the project's scale and scope (e.g. per total floor area and project type-shopping mall, residential, and more- which data itself).
- Finally, extraction of meaningful data and information

For instance, assuming the office used and trashed a ton of A4 papers from design to completion of a 1500 square meter floor areas shopping mall. Then, a useful data/information is "1 metric tons of paper wasted per 1500square meter floor area shopping mall; also, 0.67 kilogram of paper per square meter floor area." Another data or information has environmental implications. For example, if producing A4 paper consumes 2-13 litre of clean water (as in [49]) and 52 Kilojoules of electricity in 2015 (as in [50]). Then, the hidden environmental cost of 1 metric ton of A4 paper equals: 0.4–2.6 millions liters of clean water and 10.4 million Kilojoules of electricity to produce 200000 A4 papers as trash). Further, (see Appendix 1) complex data table can be developed by substituting, one at a time, paper with electricity, gasoline, diesel, water, ink, timber off cuts, and more. Comprehensive and AI-Blockchain integrated smart table can predict, recommend, and support decisions, including climate change mitigation strategies. For example, in [51] was business case for Blockchain Carbon trading.

Methodology:

The architect's systematic process is with a loop of thinking, making (build detailed and engaging models), breaking (open to criticism) and repeating (gathering feedback) [52]. Like all sciences, architecture has laws, principles, and theories guiding actions including the manipulations of *numeric and categorial data, structured and unstructured data.* For example:

- Principles include: balance, affordance, feedback, mapping
- Theories: form follow function; architectural design as a social process.
- Unstructured data like surveyor site plan dimensions, client's demographic details (client age, income, and family size), site's vegetation, topographic detail (slope), and ethnographic data like cultural symbols characters. Including narratives are analyzed, clean, labeled, structured, and stored securely for easy retrieval and use.
- Structured *data are clustered* into compatibility, noise area, private areas, and semi- private areas (i.e. data modeled overriding some significance issues).
- Synthesis using real data: models including bubble diagrams, zoning, 3D computer model, physical model, figure. Update data base which indicates the use of "*real data:*" including the use of streaming data to analyze traffic around project site using *modern technologies* like Google Earth, Drones.

Supporting the claim of the science and statistic skills demanded of the architects is in [53, 54]. The use SPSS, graph and reports, of *z-score, t-score, normal distribution curve* to generalize findings by architects is common. Architects perform *regression analysis* to determine relationship between architectural productions and users' demographics. Indeed, scholars disagree architecture is more complex than machines because humans are not machines. These are useful for decision making [54, 55]: as consultant, public policy maker. For example,

- Household income and house typographic relationship; it is important in housing studies and housing projects.

- Also, regression analysis is used to determine household income, age, or occupation relationship with transit choices (car owner, mass transit, and pedestrian behavior); important for discerning group behavior and infrastructure use patterns; informed sustainable transit infrastructures spatiality in neighborhoods.

- Regression analysis can also reveal motorist most probable behavior when relative to different stress levels such as in traffic, argument, extortions, and noise. It is important in giving insights and determining standards for walkway design, strength and proximity to road, including safety in spaces-places.

Finally, the architect is *open to criticism, creative, team player, analytical, numerate*, with products aimed at making *impact in society*. The data gathered and generated from the above are appropriately managed and stored in the cloud (Google, IBM provide cloud storage services), Dropbox, and alike for easy access and use, including updating data outside the office. Each circle yields more data and progressively better the Design; called iterations. The point is, the architects collect the data, analyze it, label and classify it, and store it safely. He or she also retrieve and use the same for future project or training manual. All these involve using modern tools. They are consistent requirements of a data scientist [56].

3.5 The city

- Smart city

A term that has been the tonic with the political, built-up space, but most commonly the Smart city is understood to mean only a sophisticated computer driven city or even a "laboratory" in [57]. Conversely, other scholars think smart city is not all about a computer city, but that "the value lies in how this technology is used rather how much of it is used" [58]. In support, the chapter point at the intersection of using computer to identify appropriate data and use of such data appropriately. For instance, **Figure 3** indicates smartness rather than poverty; knowing what is sustainable, available, and using it rather than using what is sophisticated, yet on loan, and not sustainable. In other word, to deliver responsiveness in demand, justice, sense of belonging and wellbeing among city users in addition to availability of internet is smartness. In cities, it implies shared risk, wealth, collective decision making were the individual differences of the majority are significantly respected, rather than being over ruled by a few. In [58] data availability, integrity, confidentiality, accountability and operation laws were identified as canonical. Indeed, these requirements are significantly possible by use of AI-Blockchain technologies which is inspiring the big data sciences. As in [5] so, cities' safety and security can be preserved from detrimental

Figure 3
Ijaw stilt house: smart and simple.

behaviors by data analytics of appropriate big data (extracts of terrorists, cells, deviants). City's infrastructures can be more responsive and preserved by citizens when infrastructures associated decisions, policy and laws are democratic in origin [59].

Poor data can impact cities negatively as earlier demonstrated. With disadvantageous data, subjectivation of the city: (a) cities abating crimes, (b) assume wrong identity, (c) anti-citizenship. The fact, AI-Blockchain can also learn the wrong stuff, if errors are not immediately corrected. In that sense, no matter the level of computer sophistications such cities are not "smart". Increasingly as symbols of exploitation, and battlefields of the winner take it all. Yet worsen by financialization and capitalism. Indeed, worst is the thinking of cities as "laboratory" which may justify why extremist as leaders can imagine turning cities into "radioactive ash" [60].

- Leadership Failure and Impact.

Considering architects' big data capacity, so why the disillusionment? Poor motivation. In a conversation the authors had with scholars of the Port Harcourt Polytechnic O. Wodike, G. Irighmah, C. Chendu, F. Dimabo blamed client's unwillingness to financially commit adequately for architectural services as reason for lack "focused research" required for "every project is unique." Thus, the attendant weak "architectural language" development, and repetitive narrative in cities. Besides, clients do not exclude the government while the "digital transformation is no longer optional" [61]. Therefore, poor data in addition to fragmented data culture in Nigeria's construction industry has enabled the impunity and bad image of the sector. In [62] Executive Director of Project in the Niger Delta Development Commission (NDDC) created to provide infrastructural gap in the Niger Delta region, described winning NNDC contract as a winning lottery to emphasize the impunity in the sector. Politicians have hijacked the data management weakness and unrestricted entry.

Today, politicians and public officials seamlessly inflate building and construction projects with a 2021 survey reported 56,000 abandoned projects valued about N12trillion, scattered across Nigerian cities as a result [63]. The construction sector's porosity, simulating a zero knowledge-based sector, has enable misfits into the sector's strategic leadership. Unlike with the health (Federal Ministry of Health), and justice sectors (Federal Ministry of Justice) that allow only doctors and lawyers as ministers and commissioners to head the ministries. In Nigeria, the Ministry of Housing and

Works, Ministry of Transport, and Niger Delta with huge infrastructural budget have lawyers as ministers with so much power. Expectedly, the misfit may account for the huge "debt implication", "lambasting,' politicizing human right (e.g. right to descent shelter), and corruption [64–66] with impunity. A misfit issue becomes important, with a disadvantageous data management system yet securing infrastructure financing loans from the likes of China [67]. Thus, the demanding leadership and accountability from the construction industry leaders as in **Video** [68].

More details of poor performance:

The Ministry of Transport construction of a USD 1.9 billion standard gauge railway from Kano to the Niger Republic is causing uproar in the Senate, including the citizens [66]. Relative to the dreadful roads plaguing the industrial and commercial hubs, including Wes Africa largest market Onitsha (South east of Nigeria), the Kano-Niger Republic railway is economically disadvantageous; indeed a symbol of marginalization and insecurity as illustrated in **Figure 3** [66]. In Guardian November 10, 2021 , the Nigerian Institute of Surveyor "Call For Subsidence Monitoring & Deformation Study of High Rise Structures" following incessant building and overpass collapses in Nigeria [63].

3.6 Possible information and telecommunication technologies for students

As matter of emphasis, it is necessary to dedicate a subsection dealing with possible IT technologies. Architects must constantly seek ways to improve their communication tools and channels to clients and all stakeholders. Thus, this subsection will offer in specific and significant detail the possible data technologies for both students and architects; now and into the future. The list of possible IT technologies cannot be exhausted as more are emerging and others evolving fast too. Below paragraph are few in use and to be use by Architecture students and architects:

Excel, Google form, Mind Mapping tools, Google Earth, others for common graphics editors are: Photoshop, Illustrator, Paint Shop Pro, CorelDRAW, cloud strorage. In addition, the Apple Photos plug-in from Macphun, Adobe Lightroom, Digital Image Suite, Canva, and related programs. For designs and animation, 3D data works are Revit, ArchiCAD, BIM, Sketchup, and Mind Mapping software (e.g. Microsoft Mesiere, Yed, see **Figure 4**). While the aforementioned are already in use, the future increasing immersion and free-hand experiences have Microsoft HoloLens 2, SketchUpViewer, demonstrated in Vieo-6. Blockchain, NLP, 7D, 6G, and Digital Twins (DT) as IT technologies the architects will depend in everyday service delivery. A few technologies are discussed next (**Figure 5**).

- *Excel:*

In [69] is "a software program created by Microsoft that uses spreadsheets to organize numbers and data with formulas and functions," according to Corporate Finance Institute (CFI). In [70] Excel was described "as a way to create nice looking tables of data – things like building programs or drawing lists." Also, it has mathematical formulas with capacity to analyzing and computing data. ArchiDaily, in the same [70] listed twelve useful Excel formulas for architects few are: SUM, IF, SUMIF, COUNTCOUNTIF, COUNTBLANK, AVERAGE, and more. SUM will enable the performance of mathematical operations on data (i.e. parametric data). "IF" will enable architecture student perform conditional analysis, e.g. IF income is income higher than N300000 monthly group data as "middle income." SUMIF will enable addition of specific group of data if specific condition if fulfilled, e.g. SUM the number clients,

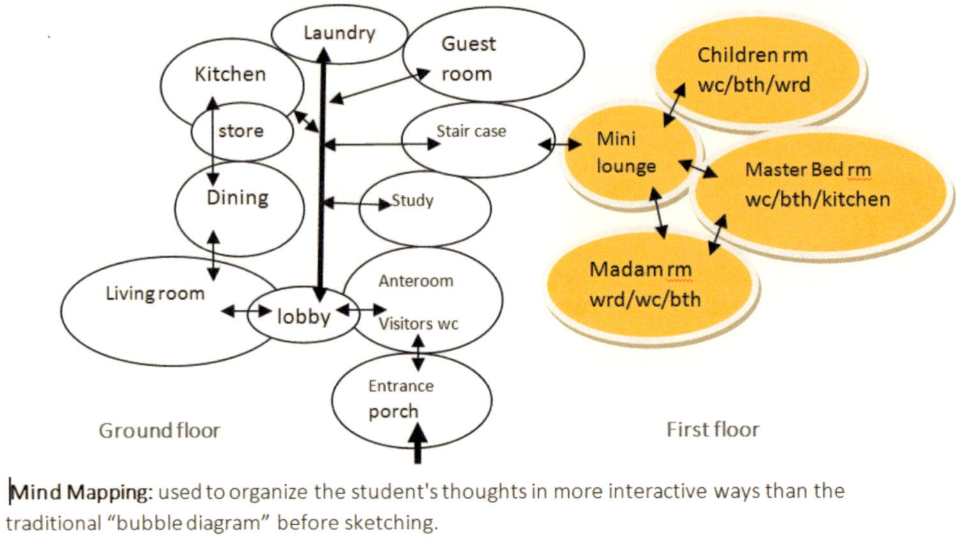

Figure 4.
Bubble diagram achieved by 3D computer graphics with arrows indicating human circulation. Credit: Anita Bala.

Figure 5.
3D graphics and Data use is serving as an essential component of computer-assisted learning system. Mix-reality, a future .But current in market at USD45000 with Microsoft HoloLens 2 which is mix- reality headset will be increasingly use for interaction.

IF income is above N300000monthly. The COUNT function counts specified element in cells, like none parametric data- "Female", "Married" and as such. In addition, Excel makes easy the non-pleasurable aspects in real estate. For instance, the management of land related complex and multiple sales demands involve keeping track of different measurable areas, associated varied facilities, areas for outright and installment purchases, and lease areas.

Kilkelly further acknowledged, Excel offered, "a lean and mean Excel spreadsheet accurately calculated the leasable area. Using the formulas I had built, we could quickly play out scenarios for our client" [70]. The point is, Excel is suitable for: Data input, Data management, Accounting, Financial analysis and analytics, Charting and graphing, Programming, Time management, and Task management (as in [69]).

- Tablet and Digital Art

Other than desktop, Laptop, powerful tablets have emerged to take make digital arts easy and handy. Apple Pencil, Gamon PD1560 15.6, Wacom Cintiq 22, Xp-Pen Artist Display 22R Pro, HuionKamvas Pro-Pro 22, while in [71] listed top ten "architecture apps (that) have completely altered the working habits of architects" to include: (a) DroneDeploy "developed to care to the growing need for aerial site intelligent software," (b) "ArchiSnapper is a field reporting App that allows users to efficiently draft a report during site visits- users can take photo, draft and comment and the app generates seamlessly the report,"[71] others are Shapr3D(IOS), and RoomScan Pro. The 3D modeling process produces a digital object of value that can be sold, hired for fee, transferred as valuable gift. One of such object as None Fungible Token (NFT) like rear art can be sold for thousands to millions of dollars. In [72], for instance, the "Artist Beeple's Crossroad fetched $6.6 million on Nifty." Few architects are now specializing in building NFTs: owning virtual estates, including cartoon characters, and football stadium, like in the real world.

- Blockchain

Looking at HoloLens 2 and SketchUpViewer demonstration, blockchain value emerges as it can permanently keep track of all changes, clients approvals, and serve as basis for undertaking contract. Thus eliminating copyright violation issue, and providing evidence for contract violation. In future, a combination Microsoft Hololens 2, SketchUpViewer, digital twin (DT) technologies, and additive manufacturing or printing all on Blockchain would make it impossible to violate construction in terms of design and specifications. A very common trend in Nigeria: roads built without drainage, kerb removed, and asphalt thickness highly compromised.

- Photoshop, Lightroom, and Adobe Illustrator.

Architectural profession involve the use of photography which good marketing tools and skill. Students are taught photography as core course. It is important in content creation for books, magazines, gallery, and graphical objects enhancement. A picture "is a thousand word" declared wisdom; even so, the quality matters. Adobe Photoshop is one such powerful tool for high quality photo editing. The Lightroom as an editor and organizer is judged much better than most photo editing alternatives tools, including photoshop for easy and light manipulations. For example, with over 100 possible different filters settings Lightroom can improve a given poor quality image to great photograph for different themes: landscape, food, city view at night day, construction sites, presentations, and more. Nevertheless, alternatives to photoshop are many. Adobe Illustrator, according to the Graphic Design Institute's publication also facilitates creating stunning art.

3.7 Possible information technologies: the future of architecture

Looking at the future, some technologies are already in the market including payment systems. Before considering Hololens 2, that chapter will present a case that may support the invalidation of fear by architects over losing jobs to computers; the case of Web 3.0 and None Fungible Tokens (NFTs).

- None Fungible Token (NFT):

Which are digital objects of unique kind- two cannot exit. Paola Tasca, the Executive Director, UCL Centre for Blockchain Technologies in [73] recommended "tokenise business model," "tokenised economy" "tokenization of everything" [74]. Architecture students' photos can be digitized. Then, using algorithm converted into NFTs. Original outstanding architecture sketches, hand drawn inclusive, can all be tokenized. There increasing importance is not in doubt with the emerging Web3.0. Indeed, 3D objects are vital in character animation and special effects processes because they can be animated. Described as digital artwork, and "unique cryptographic token" (NFT), Sharma quoted JPMorgan analysts: "the monthly sales of digital tokens (NFT) 'hovers' at around $2 billion" [72]. In [73] recommended "tokenise business model," "tokenised economy" "tokenization of everything" [74]. "The collapse of Moore's Law" [75, 76], in addition, educational systems and business enterprises are adopting computer graphics for data visualization, graphical data processing, interactive learning, immersive learning, virtual and augmented reality, presentations" [77].

- Microsoft HoloLens-2.

Introduced in 2019, the Microsoft HoloLens-2 is a pair of mixed-reality smartglasses developed and manufactured by Microsoft, **Figure 6**. It unveils the future of the free-

Figure 6.
Microsoft HoloLens-2 introduced in Spain at the Mobile world congress in February 24, 2019 with USD3500 pre-order value. While picture speaks 1000 words, a walkthrough video speaks a thousand pictures. Credit: Anita to AJP.

hand computer. It has Bluetooth, camera, CPU, sound, and holographic lenses. The tool enables students to pull out the 3D graphics design and virtual tools from the computer into the real world or project site. With free hands -not mouse, or pen – any objects can be manipulated like in the real world. Architects, engineers, and clients can with their hands measure headroom, touch, move things to desired positions- life size or reduced scale of things. In more realistic ways, the Hololens-2 saves architects (AEC professionals) the stress of many explanations of ideas by providing the same capabilities of the real world into the virtual world (mix-reality). Microsoft demonstrated possibilities of Holocene-2 is in [78] video-6, in [79] **video** and in [68] **video 3**. Indeed, the immersion will enable reduced amend reoccurrences, step up the tempo of design, and manage customers in new ways. Further, students can identify risk earlier and accurately from design through to construction.

- Digital Twin (DT).

Yet another interesting and possible IT technology for architectural uses is DT; which enables architects, students, and schools to collaborate like never before in real time, with real world experiences. In [80] DT will enable sharing the power of super computers hosted in cloud as algorithm; used for complex computation and visual simulations. For instance, building performance in earthquake, flood, and high genetic impact: such as accident or terrorist attacks. Also, in additive printing with powerful sensors, DT can enable changes in CAD drawings to be automatically updated and reflected physically at construction site. Today, such is not possible as changes in CAD drawing may take days and weeks to be implemented at construction site.

Ordinarily, a significant number of tertiary institutions of architecture and engineering in Nigeria can only dream of such super computer hardware, but with DT the narrative may be different. DT of such hardware computer is affordable yet without compromising performance quality relative to that of the corresponding hardware (physical twin). Amazing opportunities of instinctual interactions, collaboration of students and lecturers across national borders can best be imagined [81]. For example, imagine MIT, Harvard and Port Harcourt Polytechnic sharing the same equipment, faculty members in real time without need for travel. Internet of Things (IoT), Industrial internet of (IIoT), and Internet of Value (IoV) powered by powerful sensors are fore runner of DT technologies opportunities.

- SketchUpViewer.

SketchUp is architectural software like ArchiCAD, with many use savvy students. Collaboration with Microsoft to release SketchUpViewer that is compatible Holocene-2 is game changer: see video 6. So, students can draw with the computer, but also can pull-out the designs into the real world before construction- mix reality. The point is, 3D data interaction are getting better giving students and architects more free time for more creative none stereotype activities. Also, greening the process by eliminating work travel need, waste, and enabling one-to-one interactions of human-to-human, human-to-machine, including the virtual-to-reality [81].

- Educationally:

Interactive learning software, multimedia software, online courses and many other applications have received much attention. A view supported in [77] stating, "It touches many facets of everyday activities: online or offline digital content". Others like in [82] stressed play is essential to learning: "Literacy and numeracy development". The chapter recommended the use of 3D computer graphic strategy and tactics used to engage players to be used for development of instructions for architecture students; a potentially significant and educational application area. Indeed, 3D Computer Graphics and Data use is serving as an essential component of computer-assisted learning system [82]. The statistics associated with the use of the 3D data technologies are impressive. For instance, in [83] "83 % Decrease in training time. Northeastern University has reduced a 3-hour lesson plan into less than 30-minute. 50 % Better student scores Case Western students in the HoloLens lab scored 50% better on retention and required 40% less class time". Indeed, 3D data technologies can "improve learning result and revolutionize curriculum with hands on lesson plans that convey complex concepts in 3D. With HoloLens2, students can learn-by-doing from anywhere with holographic instructions and assessments" [84].

3.8 Architectural institutions and educators

Considering the fast pace of development, highlighted challenges have been discussed. A case for architects, engineers and construction (AEC) professionals having well integrated and shared data management structure has been established. A database for comparison and control projects cost inflation in public projects. Architects' duty and obligations include building liveable communities. Thus, the chapter is also interest in pollutions in neighborhoods and communities. Therefore, architectural educators should be passionate in data sciences with a goal of establishing a Construction Blockchain (see [69, 81, 85]).

It demands more openness and agility. It demands improvement of career options in architecture to attract young people interested in creative discipline (e.g. architecture) and trending high-tech jobs with good pay and prestige. While calculus, mathematics, statistics, algebra are core of data sciences in [70, 85, 86], it is very encouraging that coding and programming language like Python do not require mathematics, or even an engineering degree, but critical thinking, little play, and design principles. The authors are architect yet good in Microsoft Excel, coding, and python. Furthermore, with Additive Manufacturing and printing as the future of construction and manufacturing, including Microsoft Hololen, 7D AR and powerful sensors, it is imperative for educators and institutions to build capacities aligned to working with CAD transformative experiences and methods (see **video** [86]). Progressive and impactful institutions must embrace strategies and tactics consistent with the learning organization as supported in [87].

Building compelling strategies and tactics can include incentive (e.g. scholarships), explore and exploit full potential of data sciences skills by sensitisation. For example, making high school Data Sciences Credit pass (General Certificate Education, or West Africa School Certificate) as Physics alternate course requirement for admission into architecture in polytechnics and universities. Next, is to move people into result-oriented actions by positive influence. However, in [88] influence can be animated by making real changes across relationships. Scholars agree, influence is magnetic and can pull people into the institution's orbit with the power to make real global changes across relationships. In [89], the wisdom to overcome barriers was revealed: colleges and universities should invest in digital transformation for more accessibility, flexibility, and affordability. They must overcome tendency of being just owners of hardware equipment or DT by becoming enabler of processes delivering:

- Distinguished skill for their students,
- Advance working performance,
- Further their mission.

In [90] was the insightful demand on AEC professionals' educators. That "Construction industry: more needs to be done." For example, the importance or reliability of psychometric is yet questionable, [91] so, the use of drawing boards. Yet both are compulsory with Psychometric test pass being mandatory for graduation in Nigerian polytechnics. However, such policy is not strategic to the AEC professional development and learning experiences. For instance, many young vibrant students are discouraged seeing Psychometric course as more like an ambush and exploitations. Moreover, time is scarce for required courses. In significant number of schools in the Niger Delta, there is lack of competent staff with centralisation as a problem. For example, in some cases, much as 50 percent of students psychometric test results are delayed yet mandatory for graduation. Students have suffered additional session and associated extra fees.

Institutions should build citizenship and raise the knowledge base of the construction industry, purposefully. Construction Blockchain can provide a platform for team growth, accountability, and trust. While studios should be more conducive, collaboration is required between architectural faculty and data sciences. In [92] is the call to "upsize your strategic practices, implement new marketing, including finding new ways to build the students and faculties strategies (e.g. careers and curriculum options, lowering entry barriers for women, image improvement). As investors and decisions are increasingly influenced by big data, educators must increasingly embrace big data-driven decision making and develop competence in these areas, in addition to gradually downplaying the traditional methods. Blockchain and metaverse are good direction to prepare architects to lunch into the virtual world. The architect's creativity and expression through various media is not in doubt, **Figure** 7. Web 3.0, IoT, IoV are rather consoling and basis for confidence building among architectural educators and curriculum developers; rather than the fear of becoming obsolete. They represent opportunities and benefits from ownership of NFTs in a "tokenize economy" [93]. Architects can be significant contributors and virtual estate owners by embracing Web3.0, AI, Blockchain, and

Figure 7.
Analytical presentation and communication skills: Architecture student presentations and graphic communication through various media: Project design night Club.

animations technologies. Currently, "fintech firms are using Bitcoin blockchain and mobile technologies to create the internet of value" [94].

3.9 Conclusion and recommendation

The chapter has demonstrated similarity between the architecture and data scientists in terms of personality traits, educational skills requirements, and use of data. Architecture and its productions are scientific derivations involving multiple professionals with enormous as big data as by product. 3D computer graphics maintain symbiotic relationship with architect and architecture was explained. Further, the impact of disadvantageous data on the construction industry and the city were discussed. The fear of machines taking over the job of the architect was invalidated, because there are new opportunities in the creative economy. Indeed, there is limit as what computer can do. On career and image, Data science is a compatible as career option for young architects seeking trending high-tech and high in demand yet creative profession. Collaboration is canonical and such will add unique value to the data scientists' community. Data science can help enforce transparency, accountability and create a new wealth economy for all. Big data creation and use in the construction industry and Blockchain are critical to the desired transformation; particularly for Nigeria. The implications of policy, design, and education were also highlighted. Finally, the uncertainty demands the adoption with strategy for the organic development of the data sciences and technologies as appropriate. In Particular, Blockchain in risk free area to build capacity.

Recommendation:

- The educator must share an interest in the underlying technologies industry 4.0, web 3.0.
- Educators must understand what different levels of workers want to use the technologies in accomplishing
- Building the strategic responsiveness space distinct capability or core competence to protect the knowledge-based construction industry leadership from being open to all.
- Complete control and liability of public relevant ministry on the institute and regulatory bodies which hitherto create barriers to both imitation and mobility
- Promote synergistic cooperation among the industry professionals towards building a shared and distributed pool of big data.
- The organic development of the data sciences and related technologies like AI and Blockchain.

Finally, the imperative also Blockchain adoptions in risk free areas, as follows: (a) The legislators must demonstrate interest and political support by creating the enabling laws to give Blockchain driving big data sciences the required legal backing in Nigeria. (b) The law enforcement agencies must demonstrate interest and learn about the technologies to collaborate with the self-governing system.

Acknowledgements

Everyone who contributed the article success peer-reviewers, UNSPLASH, my wife, Peace J. Allison, for typing and Love and Unity Humanitarian Foundation non-obligatory support grant.

Conflict of interest

There is no conflict of interest, whatsoever.

Appendix 1: Survey response: evaluating the awareness and participation in data use in the construction industry. Survey by Anita Bala 2021

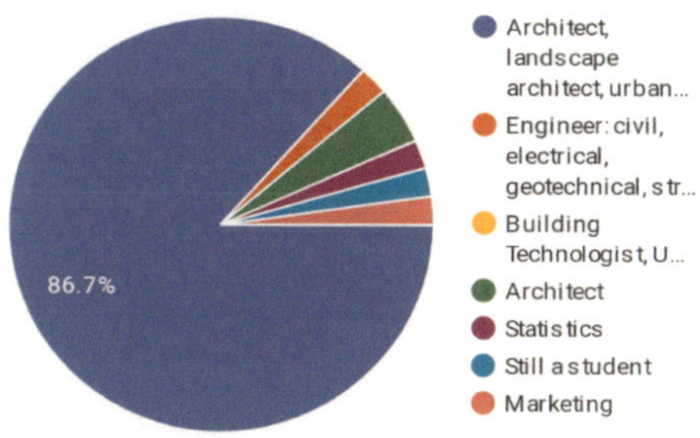

(a)

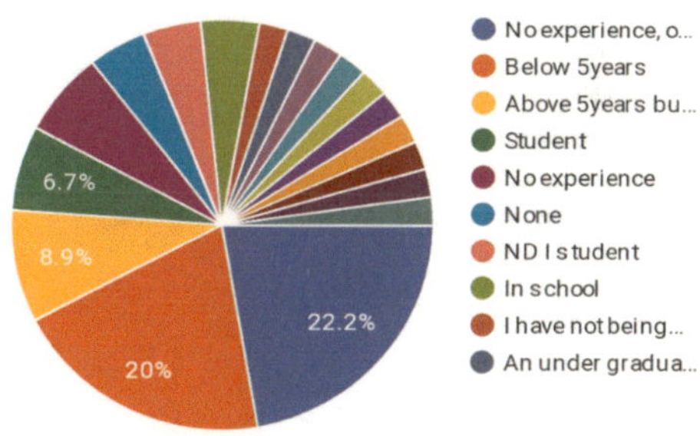

(b)

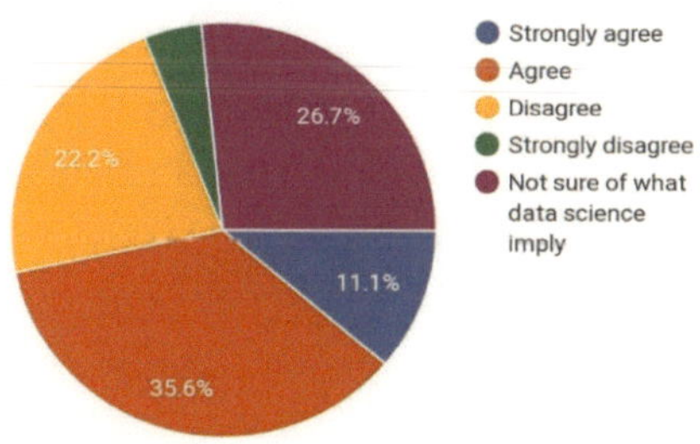

(c)

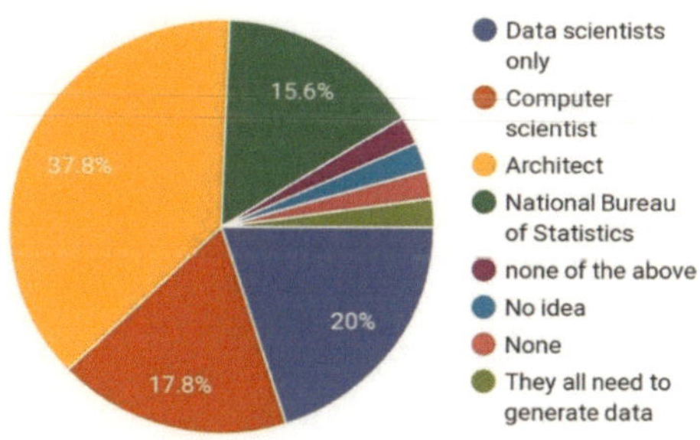

(d)

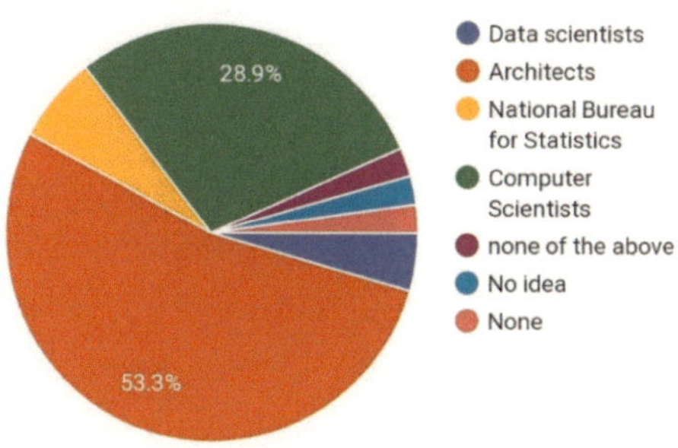

(e)

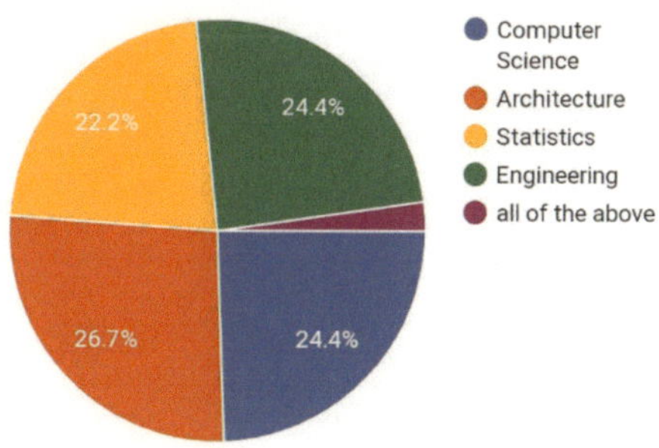

(f)

Appendix 2: Specific consumption item per iterm rate for 1500 floor area

Shopping Mall Project								**Design & tender stage/1, 500 sqm**
ITEM	**Gasoline**	**Diesel**	**Water**	**Ink**	**Timber off cuts**	**Electricity**	**Steel off cuts**	**Time**
Water	X	X		X	X	X	X	Date
Electricity,	X	X	X	X	X		X	
Gasoline,		X	X	X	X	X	X	
Diesel,	X		X	X	X	X	X	
Water,	X	X		X	X	X	X	
Ink,	X	X	X		X	X	X	
Timber off cuts,	X	X	X	X		X	X	
Steel off cuts,	X	X	X	X	X	X		
Others								
Construction stage/1500SQM								
								Time
Water								Date
Electricity								
Gasoline								
Diesel								
Water								
Ink								
Timber off cuts								
Steel off cuts								
Others								
Post construction stage/1500SQM								
								Time
Water								Date
Electricity								
Gasoline								
Diesel								
Water								
Ink								
Timber off cuts								
Steel off cuts								

Shopping Mall Project								Design & tender stage/1, 500 sqm
ITEM	**Gasoline**	**Diesel**	**Water**	**Ink**	**Timber off cuts**	**Electricity**	**Steel off cuts**	**Time**
Others								
Maintenance to scrap value stage								
								Time
Water								Date
Electricity								
Gasoline								
Diesel								
Water								
Ink								
Timber off cuts								
Steel off cuts								
Others								

Author details

John Allison[1,2]* and Anita Alaere Bala[3,4]

1 University College London CBT, University College London, London, United Kingdom

2 School of Environmental Sciences, Port Harcourt Polytechnic, Port Harcourt, Nigeria

3 Department of Architecture, Rivers State University, Port Harcourt, Nigeria

4 AKIEL Blockchain Centre, Port Harcourt, Nigeria

*Address all correspondence to: allisonj70ng@yahoo.com

References

[1] Daues JJ. Integrated design engineering and CAD/CAM application. SAE Transaction. 1987;**96**(2):1483-1493

[2] Cukier K, Mayer-Schoenberger V. The rise of big data: How it's changing the way we think about the world. Foreign Affairs. 2013;**92**(3): 28-40

[3] Barnhorn A, Caudill L, Obbie K. 60years of CAD infographic: The history of CAD since 1957. CADENAS PARTSolution. 2021. Available from: https://partsolution.com/60years-of-cad–infographic-the-history-of-cad-since-1957/ [Accessed: December 12, 2021]

[4] Cowcher K. Kongo graphic writing and other narratives of the sign. Gainesville: African Studies Quarterly. 2014;**14**(3):141-142. Available from: https://www.proquest.com/docview/1518928882

[5] Shah P. Post Covid-19 supply chain optimisation for the Indian Pharmaceutical Industry using AI Techniques. Intersect: The Sandford Journal of Science, Technology, and Society. 2021;**15**(1)

[6] Video 1 available from (can be viewed at) YouTube; 2021. Available from: https://youtu.be/oFgpnI97h8E

[7] What Makes a Great Data Scientist? Survey Summary Report, SAS Institute. 2014. (Internet). Available from: https://www.sas.com/content/dam/SAS/en_gb/image/other1/events/WMAGDS/DataScientist-survey-report-web%20FINAL.pdf

[8] Ofori G, Toor S. Leadership and construction industry development in developing countries. Journal of Construction in Developing Countries. 2012;**17**:17-21

[9] Allison J, West I. Connection and importance: Blockchain technologies, digital humanities, and architectural heritage. In: Ugwuorah AN, Aloni C, Tubobereni IF, Amadi DE, editors. Environmental and Human Development. 1st ed. Port Harcourt: School of Environmental Sciences, Captain Elechi Amadi Polytechnic; 2020. pp. 151-161. ISBN 978-978-65247-1-4

[10] ACRP RESEARCH REPORT 214: BIM Beyond Design Guidebook. The National Academy of Sciences Engineering and Medicine Transportation Research Board. Available from: https://www.nap.edu/read/25840/chapter/#1119

[11] Guide to Pencil for Young Architects and Designers, Architectural Community. Rethinking the Future. Available from: https://www.rethinkingthefuture.com/architectural-community/a5676-guideto-pencils-for-young-architects-and-designer/

[12] Jones BJ. Computer graphics: Effects of origins. Leonardo. 1990;**3**:21-30

[13] Ducker J. Graphic devices narration and navigation. Narrative. 2008;**16**:2. DOI: 10.1353/nar.0.0004

[14] Cambridge Dictionary [internet]. Available from: https://dictionary.cambridge.org/dictionary/english/graphic&ved=

[15] Data Science WIKI. 2021. Available from: https://em.m.wikipedia.org/wiki/Data_science

[16] Kitchin R, Lauriatult TP. Small data in the era of big data. Geojournal. 2015; **80**(4):463-475

[17] Jung S. Stratified Fisher's exact test and its Sample size calculation. Biometric Journal. 2014;**56**(1):129-140. DOI: 10.1002/bimj.201300048

[18] Monegain B. IBM Watson pinpoints rare form of leukaemia after doctors misdiagnosed patient. Healthcare ITNews. 2016. Available from: https://www.healthcareitnews.com/news/ibm-watsonpinpoints-rare-form-leukemia-after-doctors-misdiagonised-patient#

[19] SAS Institute. Big Data what is it and why matters. 2021. Available from: https://www.sas.com/en_us/insights/big-data/what-is-big-data.html

[20] Sun C. Why every 3d designer/architect can benefit from the creator economy in metaverse. ArchDaily. 2022. Available from: https://www.archidaily.com/974534/why-every-3ddesigner-architect-can-benefit-from-creator-economy-in-the-metaverse

[21] What makes a great data scientist? SAS Institute Inc. 2014. Available from: https://www.sas.com/content/dam/SAS/en_gb/image/other1/events/WMAGDS/DataScientist-survey-report-web%20FINAL.pdf. [Accessed 11 January 2022]

[22] Tan S. 7 personality traits of a great data scientist. Toward Data Science. Medium, 2022. Available from: https://towardsdatascience.com/the-7-personality-traits-of-a-great-data-scientist-60059873bfa9

[23] Pilgrim T. Award-Wining Paper Cuts Deep into Eye. London: Brunel University; 2018

[24] Shah P. Post covid-19 supply chain optimisation for the Indian pharmaceuticals industry using AI techniques. Intersect: Stanford Journal of Science, Technology, and Society. 2021; **15**(1). pp. 1-20. Available from: https://ojs.stanford.edu/ojs/index.php/intersect/article/download/2052/1432/7871

[25] Romereo A. A Complete Overview of GPT-3 — The Largest Neural Network Ever Created. Thoughts and Theory of Artificial Intelligence, Towards Data Science. 24 May 2021. Available from: https://towardsdatascience.com/gpt-3-a-complete-overview-190232eb25fd

[26] Additive Manufacturing vs 3D printing. Technology, GE. 2022. Available from: https://www.ge.com/additive/additive-manufacturing/information/3d-printing

[27] Video 2. Available from: (can be viewed at): https://youtu.be/xYXbcTy5dS4

[28] Harvard University. Harvard-MIT Quantum Computing Breakthrough — We Are Entering a Completely New Part of the Quantum World. SciTechDaily, July 9 2021. Retrieved 11 March 2022. Available from: https://scitechdaily.com/harvard-mit-quantum-computing-breakthrough-we-are-entering-a-completely-new-part-of-the-quantum-world/

[29] Letzter R. Chinese research to send an "uncrackable" quantum message to space, Live. Science. 2021, February 01. Retrieved 11 March 2022. Available from: https://www.livescience.com/super-secure-quantum-messagesheaded-to-space.html

[30] Video 3. Available from: (can be viewed at): https://https://youtu.be/3sSps2zUpwM

[31] Johnson A. The slow and incremental "Revolution". Journal of Architectural Education. 1984;**56**(2):49-54

[32] Tobias M. Roles and responsibilities of architects in construction projects. NY Engineers. 2019. Available from:

https://www.ny=enginees.com/blog/architects-in-construction-projects]

[33] Team Steppingblocks. Top 5 hardest working college majors [Internet]. Available from: https://blog.steppingblocks.com/ top-5-hardest-working-college-majors] [Accessed: 2022, January 12]

[34] The true meaning of the word architect. NEDC Design and Construction. [Internet] Available from: https://www.nedesignbuild.com/meaning-of-architect/ [Accessed 2022, January12]

[35] Siry J. Carson Pirie, Scott: Louis Sullivan and the Chicago Department Store. University of Chicago Press. Google Books. Available from: https://books.google.com.ng/books?id=zLKapSs3sksC&pg=PA281&Ipg=

[36] What to Study: Urban Planning or Architecture? Utep Connect Extended University, 2019, February. Retrieved 11 March 2022, Available from: https://www.utep.edu/extendeduniversity/utepconnect/blog/february-2019/what-to-study-urban-planning-or-architecture.html

[37] Guo S, Wang J. Profit distribution in IPD projects based on weight fuzzy cooperative games. Journal of Civil Engineering and Management. 2022;**28** (1):68–80. DOI: 10.3846/jcem.2021.16156

[38] Ernst N. Neufert Architects' Data. Fourth Edition. [Internet] Available from: https://pdfroom.com/books/neufert-architects-data-fourth-edition/PkdNLyN32Xr [Accessed 2022, January 12]

[39] Lynch P. 6 Designs by and for architects that made Times's 25 Invention of the year. ArchiDaily. 2015. Available from: https://www.archidaily.com/777797/6-designs-by-and-for-architectsthat-made-times-25-invention-of-the-year/5655c512e58eceb25f00

[40] Wile R. This small German Town is home to the most beautiful hydroelectric plant we have ever seen. Insider Business Insider. 2012. Available from: https://www.businessinsider.com/this-isthe-most-beautiful-hydroelectric-plant-weve-ever-seen-2012-6

[41] Allen S. The digital complex. Log. 2005;**5**:93-99. Available from: https://www.jstor.org/stable/41765037

[42] Deconstructivism. Design Buildings Wiki. 2020. Available from: https://www.designingbuildings.co.uk/wiki/Deconstructivism [Accessed: 2022, January 2]

[43] Licklider JCR. Man-computer symbiosis. IRE Transactions on human factors in electronics. 1960; HFE-1: p.4-11. Available from: https://groups.mit.edu/medg/people/psz/Licklider.html

[44] Libic J. Why Design thinking: A non-linear process. UX Collective. 2019. Available from: https://uxdesign.cc/why-design-thinking-a-non-linear-process-60d12ac6a427

[45] Bowlby C. The Palace of Shame That Makes China Angry. Beijing: BBC News; 2015. Available from: https://www.bbc.com/news/magazine-30810596

[46] Tasca P. Internet of Value: Risky necessity. In: Vadgama N, Xu J, Tasca P, editors. Enabling the Internet of Value. Future of Business and Finance. Cham: Springer. DOI: 10.1007//978-3-030-78184-2_2

[47] Shukman H. Ranked: The most difficult majors in America. The Tab. 13 March 2017. Available from: https://the

tab.com/us/2017/03/13/hardest-major-62699?itm_source=parsely-api

[48] 2020 Global Status Report for Buildings and Construction. UNEP https://wedocs.unep.org/bitstream/handle/20.500.11822/34572/GSR_ES.pdf&ved=

[49] Van Oel PR, Hoekstra AY. Towards quantification of the water footprint of paper: A first estimate of its consumptive component. Water Resource Manage. 2012;**26**:733-749

[50] Paper production energy consumption 1965-2015. Statista Research Department. 2016. Available from: https://www.statista.com/statistics/713287/energy-consumed-by-paper-production/

[51] Richardson A, Xu J. Carbon Trading with Blockchain. In: Pardalos P, Kotsireas I, Guo Y, Knottenbelt W, editors. Mathematical Research for Blockchain Economy. Springer, Cham: Springer Proceedings in Business and Economics; 2020. DOI: 10.1007/978-3-030-53356-4_7

[52] Tomitsch M. Innovation through Design: Think, Make, Break, Repeat. Cousera: Class Instructional Material; 2020. Available from: https://www.cousera.org/learn/innovationthrough-design

[53] Aburanmadan R, Trillo C. Applying design science approach to architectural design development. Frontier of Architectural Research. 2020;**9**(1): 216-235. DOI: 10.1016/j.foar.2019.07.008

[54] Deng Z. A statistical view of architecture design [PhD Thesis]. University of California Santa Barbara; 2017. Available from: http://cseweb.ucsd.edu/~tullsen/samos12.pdf

[55] Ozturk OB, Basar E. Multiple linear regression analysis and artificial neural networkbased decision support system for energy efficiency in shipping. Ocean Engineering. 2022;**234**. DOI: 10.1016/j.oceaneng.2021.110209

[56] 9 Must-Have Skills You Need to Become a Data Scientist, Updated. KDnugget. Available from: https://www.kdnuggets.com/2018/05/simplilearn-9-must-have-skills-data-scientist.html

[57] Kashef M, Visvizi A, Troisi O. Smart city as a smart service system: Human-computer interaction and smart city surveillance systems. Computers in Human Behaviour. 2021;**124**. DOI: 10.1016/jchb.2021.106923

[58] The Welding Institute. What is a Smart City? – Definition and Examples. Technical Knowledge, TWI 2022. Available from: https://www.twi-global.com/technical-knowledge/faqs/what-is-a-smart-city. [Accessed: 2022 January 12]

[59] Allison J, Allison M, Amadi D, E. Anti-democratic spaces and impoverishment: Role of roads in low-income residential areas. Nakhara Journal of Environmental Design and Planning. 2019;**16**:15-32

[60] Miller SA and Golding B. Russian news head: We can bomb US into "radioactive ash." News, New York Post. 2014. Available from: https://nypost.com2014/03/17/russian-newsagency-head-we-can-bomb-the-us-into-radio active-ash/

[61] Reiss R. After Q1 2020, Digital transformation is no longer an option. Forbes. 2020. Available from: https://www.forbes.com/sites/robertreiss/2020/03/30/after-q1-2020-digital transforma tion-is-no-longer-an-option/

[62] Ebiri K. NDDC vows to stop indiscriminate award of contract. The

Guardian. 2020. Available from: https://guardian.ng/news/nddc-vows-to-indiscrinimate-award-of-contract

[63] Editorial. Nigeria's 56000 abandoned projects. Vanguard. 2021. Available from: https://www.vanguardngr.com/2021/12/nigerias-56000-projects

[64] Hurley J, Morris S, Portelance G. Examining the debt implications of the Belt and Road Initiative from a policy perspective. Journal of Infrastructure, Policy and Development. 2019;**3**(1): 139-175

[65] Adeleke J. Corruption in the private sector: How bad? What response? In: Aderinwale A, editor. Corruption, Accountability and Transparency for Sustainable Development. Ota, Nigeria: African Leadership Forum; 2003. pp. 205-210

[66] Iroanusi QE. Senator lambast Amaechi over Kano-Niger Republic Rail Project. Premium Times. 2021. Available from: https://www.premiumtimesng.com/news/top-news/493471senator-lambast-amaechi-over-kano-niger-republic-rail-project.html

[67] Bradshare K, Nossiter A. In Nigeria China investment with a downside. The China Factor Part 5,The New York Times: 2015, December 5. Analysing Nigeria-China Trade: 2015. Available from: https://www.nytimes.com/2015/12/06/business/international/in-nigeria-chinese-investment-comes-with-a-downside.html

[68] Video 4. Available (can be seen at): https://youtu.be/LiMpeMu2ZGc

[69] Excel Definition. Online: Corporate Finance Institute. 2022. Available from: https://corporatefinanceinstitute.com/resources/excel/study/excel-definition-overview/

[70] Kilkelly M. 12 Excel Formulas Every Architect Should Know. Online. ArchiDaily; 2015. Available from: https://www.archdaily.com/632855/12-excel-formulas-every-architect-should-know

[71] Harrouk C. The Top 10apps for Architecture. Online: ArchiDaily. 2020. Available from: https://www.archdaily.com/896021/the-top-apps-for-architets

[72] Tasca P. Token-Based Business Models. Disrupting Finance, 135-148. DOI: 10.1007/978-3-030-02330-0_9

[73] Tasca P. Token-Based Business Models. Disrupting Finance. 2018:135-148. DOI: 10.1007/978-3-030-02330-0_9

[74] Tasca P. Tokenization of World Economy, UCL CBT. 2019. Available from: https://www.paoltasca.com/wpcontent/uploads/2020/04/Tasca_Blockchainlive.pdf

[75] Chojecki P. Moore's law is dead. Towards Data Science, Medium: 2019, February 19. Available from: https://towardsdatascience.com/moores-law-is-dead-678119754571

[76] Chojecki P. Moore's Law is Dead. Towards Data Science. 2019. Available from: https://www.towardsdatascience.com/moores-law-is-dead-78119754571

[77] Bhattacharya B, Chelladurai J. Computer graphics applications in the education process. The CTE Journal. **6**(2): 1-8. Available from: https://www.thectejournal.com/uploads/ 1/0/6/8/10686931/bajracharya_computer_graphics_applications.pdf

[78] Video-6. Available from: https://youtu.be/T8QWjO6eY1A

[79] Video-5. Available from: https://youtu.be/5BUUd_wVKhc

[80] Nincarean D, Alia M, Halim N, & Rahman M. Mobile Augmented Reality: The Potential for Education. Procedia - Social And Behavioral Sciences. 2013; **103**:657-664. DOI: 10.1016/j.sbspro.2013.10.385

[81] Parmentola A, Perillo A. Tutore I. et al. Is Blockchain able to enhance environmental sustainability? A systematic review and research agenda from perspective of Sustainable Development Goals (SDGs). Business Strategy and the Environment. 2021;**31**(1):194-217. DOI: 10.1002/bse.2882

[82] The importance of play in children's learning and development. StartingBlocks.gov.au. Australian Children Education and Care Quality Authority (ACECQA). 2021. Available from: https://www.startingblocks.gov.au/other-resources/factsheets/the-importance-of-play-in-children-s-learning-and-development/

[83] Rieber LP, Smith L, Noah D. The value of serious play. Educational Technology. 1998;**38**(6):29-37. Available from: http://www.jstor.org/stable/44428495

[84] Education. Microsoft HoloLens 2, for precise, efficient hands-free work. Microsoft 2022. Available from: https://www.microsoft.com/en-hk/hololens

[85] Data Engineering and Analytic. [Internet]. 2022. Available from: https://cloud.google.com/training/ dataenginee ring-and-analytics [Accessed: January 10]

[86] Video 5 available. [Internet]. Available from: https://youtu.be.com/shorts/sgnv594RyUs?feature=share [Accessed: 2022 January 1]

[87] Ellinger AD, Ellinger AE, Yang B, Howton SW. The Relationship Between the Learning Organization Concept and Firm' Financial Performance: An Empirical Assessment. A Wiley Company Human Resources Development Quarterly. 2002;**13**(1):5-22

[88] Naidu Y. Power Play: Game Changing Influence Strategies for Leaders. Milton, QLD: John Wiley & Sons, Inc; 2016. DOI: 10.4324/9781315599755

[89] Fahey K. How education can overcome barriers to digital transformation. Educause Review. 2021. Available from: https://er.educause.edu/articles/sponsored/2021/10/how-highe reducation-can-overcome-barriers-to-digital-transformation

[90] Kulkarni GK. Construction industry: More needs to be done. Indian Journal of Occupational and Environmental Medicine. 2007;**11**(1):1-2. DOI: 10.4103/0019-5278.32455

[91] Peter JP. Reliability: A review of psychometric basics and recent marketing practices. Journal of Marketing. 2014;**38**(1):1-28. Available from: https://www.jstor.org/stable/26554866

[92] Maslan A. Scale or Fail: How to Build Your Dream Team, Explode Your Growth, and Let Your Business Soar. Newark: John Wiley & Sons Inc.; 2018. DOI: 10.4324/9781315599755

[93] Tasca P. Token-based business models. In: Lynn T, Mooney J, Rosati P, Cummins M, editors. Distributing Finance. Charm: Palgrave Pivot; 2019

[94] Skinner C. ValueWeb: How fintech firms are using bitcoin blockchain and mobile technologies to create the internet of value. Springer, Marshall Cavendish International Asia Pte Ltd. 2016: p. 425. [internet]. Available from: https://book.google.com/books/about/ValueWeb.html?id=_-bVCwAAQBAJ

Chapter 8

Enabling a 3-D Cyberspace Experience Online

Bruce Campbell

Abstract

The allure of a 3-D cyberspace grows again as communication channels connect a growing number of virtual reality (VR) head mounted displays and interaction-enabling controllers. Computer graphics enable 3-D cyberspace as much as any component critical to providing satisfactory experiences. Historically, the usefulness of 2-D interfaces entrenched when developers demonstrated the feasibility of 3-D cyberspace. Now a certain amount of prolonged inertia seems necessary to promote 3-D graphics to the participation level of 2-D graphics. The HTML 5 stack and other technologies enable 3-D integration into our planetary shared cyberspace. This chapter provides a historical perspective of one researcher's methods used to provide 3-D web experiences, the lack of critical mass for their use in the past, and how the latest virtual reality integration trajectories will bring a critical mass to the 3-D cyberspace of the future.

Keywords: 3-D graphics, cyberspace, virtual reality

1. Introduction

For many 1990s-era computer science students, the science fiction book *Snow Crash* [1] by N. Stephenson suggested a powerful use case for 3-D computer graphics to advance an alternative online cyberspace whereby large groups of human beings could meet each other online in an audiovisual context that reduced the geographical and temporal limitations of the physical world.

Around the time I read the book, I sent 66 electronic mail messages to academic researchers at US-based universities who publicly announced they did virtual reality research formally. Thirty-seven of those recipients replied enthusiastically with details on their research agendas, hardware and software infrastructures, and thematic aspects of VR they intended to explore. Two suggested they had immediate funding available if I could get accepted to their campus' PhD program. Six responders suggested I should pursue working with Fred Brooks at the University of North Carolina. Six responders suggested I pursue working with Tom Furness [5] at the University of Washington. Soon thereafter I realigned my life to take the latter recommendation seriously. I wasn't the first to imagine the potential of VR to improve quality of life for those who would be willing to immerse themselves in 3-D cyberspace.

At the time I landed on the University of Washington campus in Seattle, the Human Interface Technology Laboratory, affectionately known as the HIT Lab, contained a remarkable amount and variety of VR-enabling software and hardware [2]. The HIT lab also employed a highly-effective cybrarian, Toni Emerson, who performed the lead editor role of the USENET group sci.virtual.worlds, which grew to have an avid international readership. Researchers around the world could be read posting VR-related ideas and news at any moment during any 24 h day. As a result, an endless stream of VR-based conversation and news flowed through the HIT Lab. The simple text correspondence built a collegial sense of community.

The HIT Lab worked on research grants with industrial, governmental, and academic sources of funding. The work those grants funded varied from building VR-related hardware, to building software, to performing design and development work in virtual world creation, to testing the effects of VR on human beings immersed in a wide variety of virtual experiences for extended periods of time. Funders seemed motivated to visit the lab often. I understood why. The experiences provided within immersed computer graphics were thought provoking. The human energy level in our 10,000 square foot lab space was high 24/7.

A significant change in career path occurred for some of us when we veered away from PhD work in computer graphics hardware and algorithms, to focus more on the coupling of computing platforms with human perception and cognition. Dr. Furness convinced us that the lab benefitted from being firmly located within an Industrial Engineering department, as not enough people were working on the coupling of the fruits of computer science research with the full capabilities of the human being.

Many of us in the lab decided to focus on VR as a tightly-coupled system between man and machine. By the time we defended our doctoral theses, our department had been renamed as Industrial and Systems Engineering. I had spent so much time applying VR to social spaces and investigating data streams coming from natural systems that the addition of systems to the name alleviated concerns about feeling like a fraud or imposter in the industrial world.

Whenever I walked into the lab to start a long work shift, I could not help but have a nagging thought that the HIT Lab experience deserved to be expanded to avail the experience to more people. The vision of what we were working on seemed to suggest we ourselves should be working and playing in immersed 3-D cyberspace as we worked on VR. An undercurrent of ethical arguments emerged as thoughts from time to time, as well as a burning desire to make greater access possible because it seemed technically possible—and because even more books, movies, and radio broadcasts were telling stories within that context.

Work for my masters thesis in computer science, entitled "3D Collaborative Multiuser Worlds for the Internet" [3], provided me ample opportunity to spend time in the various online VR platforms—a subset referred to as *desktop VR*—where I virtually walked about in curious visual spaces with likeminded researchers who felt VR should enable widespread visual 3-D human communications, unlimited by geographical location or temporal time zone.

Because the various platforms we met in existed in the labs of research organizations, the quality of communications was high. Those of us meeting in there, from all over the world, were building a 3-D cyberspace and imagining all it could become. Twenty-five years later, what we imagined is far from being mainstream, but the technology to build it has continued to improve and evolve.

2. Building 3-D cyberspace in the 1990s

I had leveraged a paper I published entitled "VRML As a Superset of HTML: An Approach to Consolidation" to first gain access to researchers at the HIT Lab [4]. That paper suggested that 3-D cyberspace might be a useful veneer for organizing all the 2-D content that was rapidly being amassed and distributed in the early years of the World Wide Web. The lab investigated that opportunity from a high-tech perspective, but many of us there shared a vision regarding lower cost implementations.

In 1994, the HIT Lab collaborated with a Fujitsu research lab to build a shared 3-D cyberspace, named GreenSpace, shared by participants meeting there while physically in Seattle and Tokyo [5]. A Silicon Graphics Onyx machine hosted that cyberspace's visual content and processed the various peripherals that immersed each participant in VR. Four dedicated ISDN phone lines passed voice and data packets between the two locations on opposite sides of the Pacific to create a shared presence. Open Inventor software facilitated the building and management of the cyberspace experience. The lab celebrated upon demonstrating that two groups of people could share an experience in 3-D cyberspace with a million dollars' investment of technology.

A year later, eight members of the HIT Lab got together to work on a version of GreenSpace that could run on Pentium 2 Intel personal computers, using a graphics accelerator board, and the emergent World Wide Web. The Industrial Technology Research Institute (ITRI) of Taiwan provided the funding, the graphics boards, and a couple of highly-capable collaborators to work with us to find a common ground culture for the distributed cyberspace we created. Our team comprised of two computer scientists, an information scientist, an artist, three virtual world designers, and an architect who eventually married an architect on the GreenSpace team.

We worked with Pentium 2 hardware early in the project but soon got the desktop VR experiences running on Sun Microsystems, Digital Equipment Corp, and other UNIX-flavored machines, thanks to Java 3D software and its capable Java virtual machine that made cross-platform applications easier to develop. The first experience we created on the Virtual Playground framework was a virtual outdoor mall that included embedded web browsers that streamed 2-D and 3-D content on virtual billboards.

The billboards could render most web pages that were HTML-driven, as well as show VRML-based 3-D models and examples of video on some of the underlying hardware flavors. We named that virtual world Netgate Mall and spent significant time discussing what culture we would like to enable for visitors who came to visit through Internet communication channels [6].

Upon sharing Netgate Mall with our research partners in Taiwan, an artist collaborator there built a Taiwanese version of the visual architecture for the mall that shared the same bounding boxes for navigation pathways (**Figure 1**). As had been demonstrated with GreenSpace, 3-D cyberspace on the desktop could be configurable, and thus personalized, without degrading human communications for many applications. We used the Taiwanese version to run a shared cyberspace world for 10–14 year-old children to use in Taiwan, with a general public access point in a Kaohsiung museum for those who did not have access to the hardware and bandwidth otherwise.

Sun Microsystems provided a phone hotline to us so we could call them anytime with questions or feature requests related to developing and debugging Java 3D-based applications. Within 18 months, we had used Java 3D and a framework of core Virtual Playground modules to create a virtual cadaver lab [7], virtual watershed world [8],

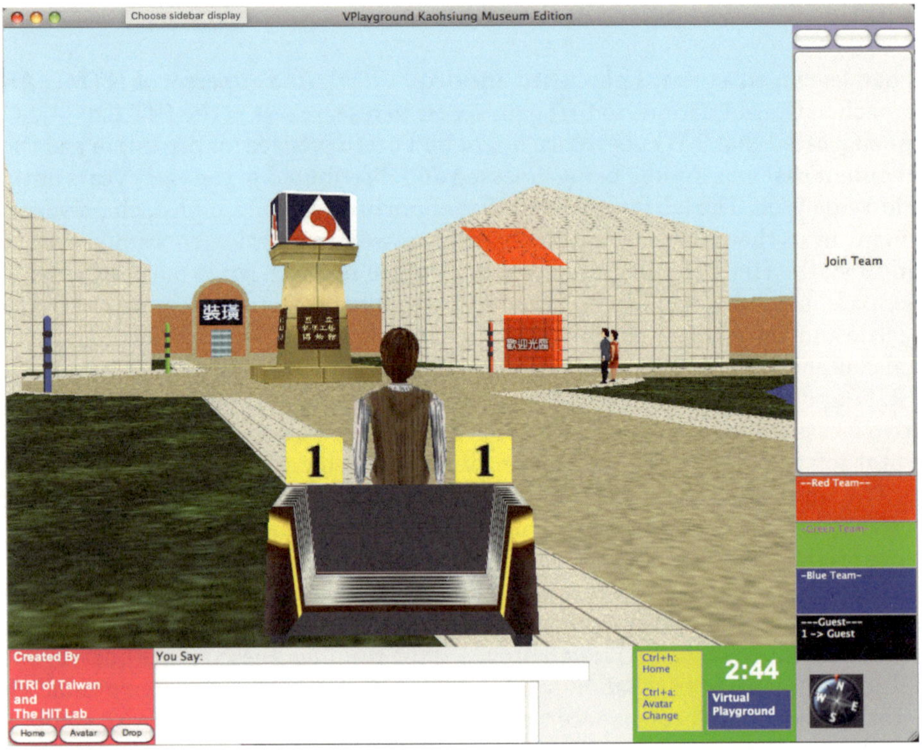

Figure 1.
Netgate Mall Taiwanese version based on Virtual Playground software.

and to integrate our applications to comply with an open 3-D cyberspace interaction specification developed through a California-based creative commons (**Figure 2**). A computer science undergraduate student ported much of the Virtual Playground GUI API architecture from Java to C++ to create BlueSpace, which then let him extend the head's up display and terrain engine.

Digitalspace hosted a workshop of the third annual digital biota conference [9] for attendees who were interested in standardizing the usability methods of desktop VR and messaging that coordinated the virtual experience between 3-D cyberspace participants. We adapted our flavors of 3-D cyberspace technologies to validate the specification and focus the creative commons on expanding the project base in which the specification could be verified. We spent a lot of time together in 3-D cyberspace envisioning possible futures.

I worked on demonstrating an in-browser solution that could provide a desktop VR entry point for participants who had not had any 3-D content experiences. A team in Turin, Italy created a native Java engine that supported 3-D cyberspace experiences. The whole engine could be downloaded in less than 100 kilobytes to plug-in to the standard Java virtual machine. Another team outside of Melbourne, Australia created the 3-D models that conformed to the engine's content loading functionality. We developed JavaScript code that drove the user experience through an HTML skin and made the skin configurable.

The first application we made available with a web-browser based engine was a configurable classroom intended to be shared by school teachers who were thinking through ramifications of the Columbine school shooting. The funder also provided a

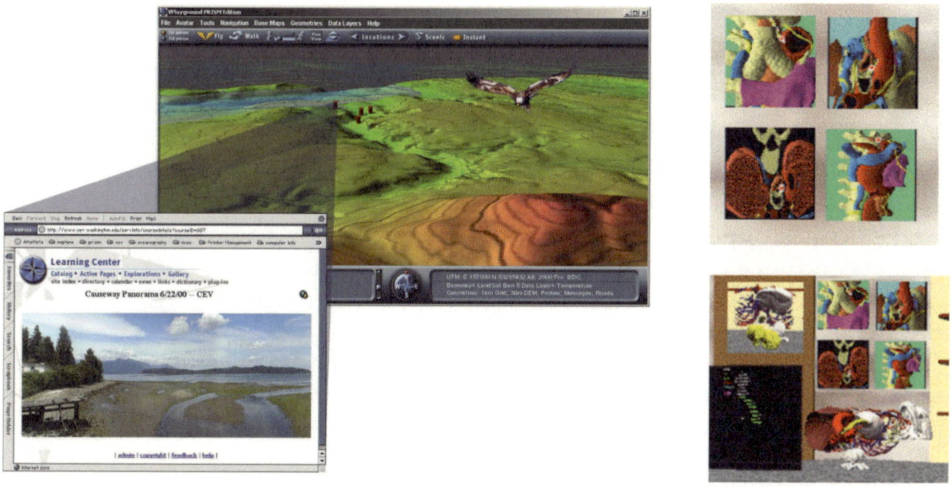

Figure 2.
Screenshots of example 1990s-era HIT Lab experiences in 3-D cyberspace. Virtual Big Beef Creek (on left) coordinated planning and geospatial organization of physical data collection (with accessible 3-D photographic perspectives). Virtual Cadaver Lab expanded a popular 2-D interactive anatomy atlas into 3-D cyberspace where self-testing was possible for learning taxonomical awareness of gross anatomy characteristics.

discussion room for school teachers to discuss their retirement investments and the insurance products provided by their employer.

The lab also maintained a virtual world created with the VRML plug-in and LiveConnect facilities provided by the Netscape web browser [10]. That world ran well enough for us to run masters' thesis experiments with four synchronous participants connected in 3-D cyberspace at a time—exploring hypotheses negotiated with an advisement committee. Soon after, Netscape lost significant market share as Microsoft's web browser took over as the leading market share browser on personal computers running the Windows operating system.

3. Building 3-D cyberspace in the 2000s

As the new millennium arrived, we had accumulated a large enough sample size of 3-D cyberspace experiences to develop a preference for specialized experiences—those that were designed for subsets of professional roles that could benefit from participating in visual environments that facilitated their roles and built community around them. We had experienced enough accumulated hours in 3-D cyberspace, though mostly by way of the desktop VR version, that every day our minds lingered there, as if in parallel lives. We concluded that creating a general Metaverse, as described in the book *Snow Crash*, required a specialty we were not comfortable in acquiring.

As a result, we formed deeper bonds with other academic departments on campus. I turned my attention to designing other useful 3-D cyberspace experiences for oceanographers. I chose oceanographers because they studied one of the two large 3-D volumes in which life thrived on the planet. The atmospheric scientists that studied the spaces where birds soared seemed to be ahead technology-wise thanks to a popular human interest in weather and weather forecasts. The 3-D world of life inside of human anatomy also seemed to be ahead technologically, thanks to the support of medical practices.

As we were contemplating 3-D virtual experiences to spur on insight regarding the nature of tsunamis and underwater earthquakes, and volcanos that can produce them, hurricane Katrina devastated the gulf coast of the United States [11]. A representative of the Federal Emergency Management Agency came to our lab to share stories of how the emergency response effort in New Orleans had been deemed suboptimal. Dr. Furness and I spent hours listening and then suggesting ideas on how VR could improve future hurricane response efforts. We began to attend emergency operations meetings at the campus, city, county, and regional level and concluded that the typical resident in a community crisis lacked situation awareness more than the emergency responders, even though they were physically based in that community.

We attended one particular meeting of many that drove that point home. A red-faced emergency operation center director, who had over 30 years of EOC experience, railed against the audacity of his county's residents in response to a severe wind event that had blown through the county just south of Seattle's. Overnight, over 250,000 households had lost power through thousands breach points on the power grid. Many linemen were working 16-hour days to restore power and yet hundreds of residents were calling in to the emergency operations center to complain that power wasn't being restored fast enough. He yelled to anyone listening that 2-week outages should have been expected, given the number of qualified workers who could repair the electric grid, and the number of homes in the dark.

Our personal experiences in 3-D cyberspace strongly suggested emergency preparedness could be a ripe place to benefit by having a thriving VR experience for residents to explore and engage in with others—including EOC personnel when available. We put together three hypotheses to test our instincts and then worked with three other advisors to whittle down our enthusiasm to a reasonable scope of experiments to perform.

We published the results in four papers [12–15] and a dissertation [16], comparing results of a physical hospital evacuation scenario drill with a virtual one. In the virtual drill, emergency responders explored the physical environment in which the evacuation was to take place by way of a software application. They could study the hospital's physical layout, transfer points to transportation, the county's road network, and 23 other hospital locations that agreed in an official memo of understanding to accept patients in an emergency.

Those in the study had access to a simulator where they could watch evacuation paths and timings for any patient in any room, based on time and motion studies done with physical evacuation efforts (**Figure 3**). They could watch simulated transport of vehicles carrying patients to destination hospitals.

A server computer communicated with the simulation clients that provided the graphical user interface for participants. The server kept track of 86 key variables on simulated patients, simulated emergency responders, transport vehicles, and events that required dynamic diversions to evacuation plans (e.g. traffic congestion). The hospital emergency response coordinator challenged the behavior of those variables as she became familiar with the client application I iterated upon in preparation of the experiments.

Adapting the server to feed a 3-D cyberspace version of the training and participation interface would be trivial and add an immaterial computational demand. Every object in the simulation was tied by gravity to a ground plane. The effort to port the participant experience to 3-D cyberspace would require significant effort in comparison.

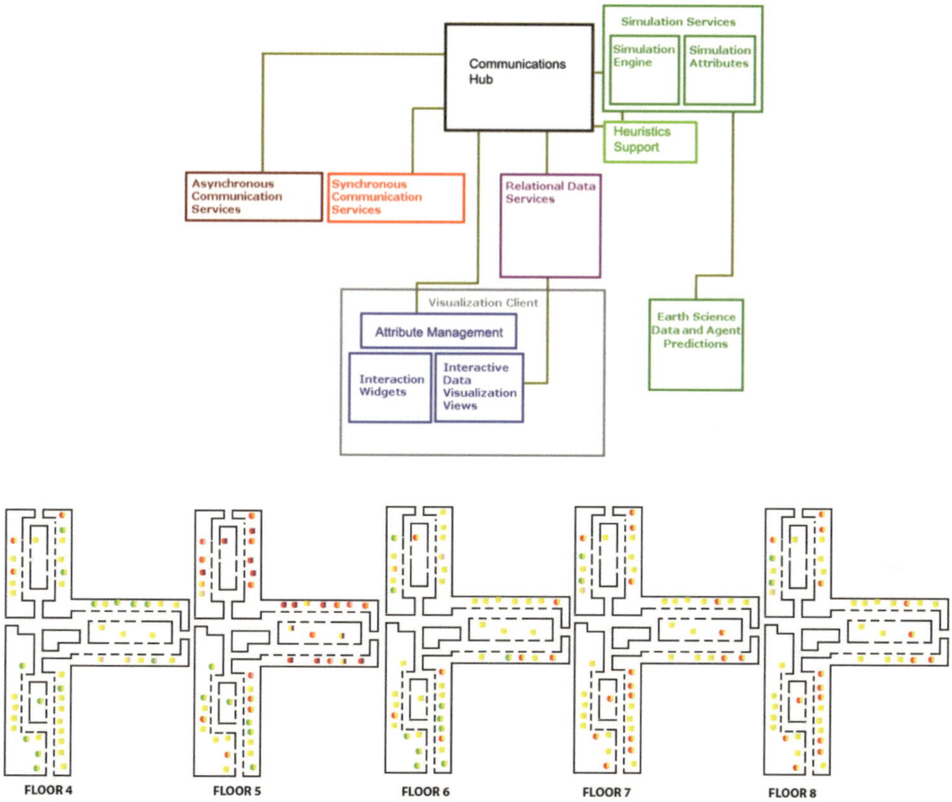

Figure 3.
The software support services for an emergency response hospital evacuation simulation application (top). Participants preferred 2-D artifacts to explore and interact with the situation as it unfolded (bottom). The 2-D hospital floor layout was used often in day-to-day operations, but patient characteristics were encoded in color for active patients from data sheets used in day-to-day operations.

Earthquake, tornado, and fire events have been included in the SimCity game application since the 1990s. Players can watch the effect of those community crises and gain insight into how urban planning matters in facilitating resilience and recovery. The third-person view of a comprehensive emergency response effort is useful. The first-person view provides an opportunity to compare a limited situation awareness in comparison—highlighting the importance of communication between all those limited views, in order to attempt to create a third-person view from first-person experiences.

Dr. Furness and I came to the conclusion that one of the roles in an emergency operation center should be a dedicated situation awareness specialist who spends much of his time in 3-D cyberspace piecing together situation awareness from the first-person reports of emergency responders—coordinated with 3-D graphical assets. Modeling and computational support would be provided by way of a coupled computing environment that contained heuristic-based equations and forecasting equations. The immersed VR role would be available to the command and control leadership in the EOC to answer questions regarding feasibility and appropriateness of actions and to evoke insight that command and control might not be considering.

I came to the conclusion that the 3-D cyberspace that would be developed for EOC personnel and emergency responders should also be made available to the general public by way of a 3-D cyberspace experience. Immersive VR might build compassion for the cognitive demands, under urgent conditions, of the first response effort. Immersive VR would give access to experiencing the delegation of roles involved, so as to prepare the public with reasonable expectations for interacting with those roles in an emergency. Overall, the immersive VR perspective would build up experiential knowledge to go with any more abstract knowledge the public has of their community.

4. Building 3-D cyberspace today

Upon revisiting the thoughts from earlier days of building 3-D cyberspace experiences in a research lab, we consider advancements in computing, development tools, and communications protocols. We see an expansion of VR peripherals into the hands, and onto the heads, of the general public. To build a 3-D cyberspace to support the Metaverse, or a local community preparedness awareness, we have many options to advance the process from the technologies we used previously.

A popular approach among our students, and the students of other teaching colleagues, for creating virtual reality experiences is by way of a pipeline approach where 3-D models are generated in freely downloadable modeling software (e.g. Blender), brought into other freely downloadable software (e.g. Unity Engine) to become part of a 3-D application, and then ported to a VR environment (e.g. Oculus) to be experienced immersively.

At the time of the hospital evacuation simulation software use, a 3-D hospital model (X3D format) was available for participants to explore within a web browser, but participants preferred the 2-D floor layouts that we provided for testing the usefulness of simulation software. Since then, the popularity of the three.js, webGL-enabled 3-D, library makes a hospital model integration easier [17]. For example, the same 3-D hospital model can be used within a Java-based application, JavaScript-based application, and Unity Engine-based application.

Figure 4 shows a 3-D model of one floor of the simulation software hospital, loaded within the Blender software in which the model is iterated upon to improve the design. The model can be exported to OBJ, GlTF, or any other file format for which the three.js library has developed a loader service to rapidly get it inside of a JavaScript-driven 3-D experience.

The Unity Engine-based application is readily made available for virtual reality use by way of XR plug-ins available from the online Unity Store. Android build support readily enables Oculus flavor VR from both Windows and Mac OS X operating systems. The Unity Engine can also import OBJ and GlTF file formats to enable interactive 3-D experiences that are OpenXR compliant. As a result, a 3-D cyberspace readily provides a virtual hospital in which to have meaningful experiences such as preparing for an emergency evacuation (**Figure 5**).

Because our resultant experience is OpenXR compliant, we could pursue extending it's usefulness through dissemination by way of other delivery formats such as A-Frame [18] and Networked A-Frame [19]. At the time of this writing, Collabora, HTC, Microsoft, Oculus, SteamVR, and Varjo provided OpenXR runtimes in which 3-D Cyberspace experiences could be delivered [20].

As we watched participants perform the virtual hospital evacuation activity, we noticed participants using the 2-D interface effectively without any apparent cognitive difficulties. All participants were given versions of the software to practice its

DOI: http://dx.doi.org/10.5772/intechopen.102416

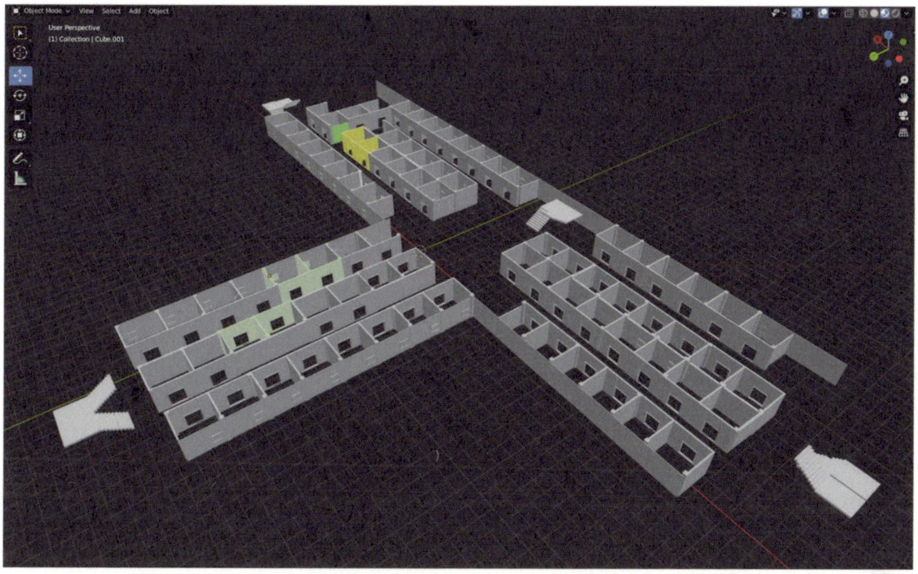

Figure 4.
One floor of the simulation's emergency evacuation hospital as contained in a 3-D model for 3-D cyberspace use, as seen within the Blender software used to embellish and maintain it.

Figure 5.
*An immersive cyberspace perspective of the 3-D model seen in **Figure 3**. Detail is faithfully rendered as part of an early iteration in preparation for use. The left image provides view into a patient room. The right image provides the view from a hallway corridor.*

use, but participants reported they only needed 2 h at most to feel comfortable as the interface was built from artifacts (maps, forms, and other documents) that participants used regularly in equivalent paper-based formats.

To gain confidence in suggesting a 3-D interface would be useful, we thought about our own time spent recording time and motion data associated with moving about in the hospital and in evacuating human beings with their necessary equipment (wheelchairs, gurneys, respirators, etc.). Having spent time doing those tasks, we could readily cognate about them with the 2-D interface. But we could not do so easily beforehand.

We forget that as humans we had to learn how to use 2-D maps as abstractions of the 3-D physical world they represent. As a result, we learn differently through 2-D abstractions than we learn interacting with the physical world. A goal in using an emergency response simulator is to improve learning with regards to the activities, context, and heuristics associated with a scenario. Some compelling evidence from neuroscience experiments suggests 3-D adds additional value even when a 2-D interface is available [21]. Conclusions reached include:

> *"A key finding is that 2-D and 3-D neuroanatomical brain models were perceived differently. 3-D learning yielded greater object recognition, perhaps from stereopsis facilitating visual recognition. Exclusive use of 3-D models should be avoided when instead teaching that includes both 2-D and 3-D is likely the most advantageous to promote generalization of knowledge by forcing students to practice fluidly between dimensions."*

Another recent study of 3-D virtual environment interfaces concluded that "findings showed different condition correlations with the traditional tasks and the comparison between the two systems have revealed that 3-D is able to generate lower reaction times, higher correct answers, and lower preservative responses in attentional abilities, inhibition control, and cognitive shifting than the 2-D condition" [22].

Common sense suggests that mental fluidity is valuable under the urgent response demands of a community-wide emergency response scenario.

5. Conclusion

We perform a thought experiment to consider the evolution of 3-D interfaces forward in time to when they would be a more dominant norm. We imagine that once 2-D interfaces no longer need to be learned for many day-to-day activities, participants would not be fluent in their use. Other large primates have an easier time using 3-D interfaces than 2-D interfaces, which seems reasonable because they have not encountered the 2-D abstractions in their day-to-day lives.

As a species we have built a cyberspace that is primarily made up of one-dimensional text and two-dimensional documents. Today's cyberspace is entrenched in part by that history and the expectation we have created that the destinations we want to visit will be 2-D at most. As a species we have also created an extensive number of 3-D models and have developed technologies to better capture physical 3-D reality in a virtual facsimile. Google Street View technology enables 3-D exploration along street locations whereby maps of our physical world provide context. Run a virtual reality application and that widely shared data provides a primitive 3-D cyberspace experience that has the promise of getting more mature rapidly. Of course large corporations continue to work on their own versions of a Metaverse using their proprietary hardware and software as well.

With a web-accessed 3-D cyberspace available, 3-D computer graphics take on additional value as a potential contribution to useful cyberspace experiences. Whether they are used for that purpose grows as a cultural question but shrinks as a technical question as a result of our progress. Exploring the cultural question rapidly expands into many sub-questions—many of which are perhaps useful to consider anyway:

1. What are we trying to accomplish as a species?

2. How do we experience our sense of community?

3. How useful is it to experience each other independent of space and time?

4. How are the physical world and virtual world co-evolving?

5. Should we enable a 3-D cyberspace first and explore what it is good for?

6. Do we build demonstration uses of 3-D cyberspace and watch as usefulness grows from specific use cases?

The potential use and reuse of computer graphics grows as the interest in connected 3-D cyberspace grows. When revisiting the emergence of the worldwide web in the early 1990s, we remember computer graphics integrating to create a cyberspace that felt like a global encyclopedia of information resources. The potential exists to revisit that time with today's 3-D technologies available so as to develop other trajectories of world wide information connectivity, with computer graphics providing a sense of continuous 3-D space in which to provide meaningful experiences. The reader is encouraged to participate in its development and exploration.

Author details

Bruce Campbell
Rhode Island School of Design, Providence, USA

*Address all correspondence to: bcampbel01@risd.edu

References

[1] Stephenson N. Snow Crash. New York: Bantam Books; 1992. ISBN: 9780553351927

[2] Human Interface Technology Laboratory [Internet]. 1996. Available from: http://www.hitl.washington.edu [Accessed: December 09, 2021]

[3] Campbell B. Collaborative Multiuser Worlds for the Internet [thesis]. Hartford: Rensselaer Polytechnic Institute; 1997

[4] Campbell B. VRML as a superset of HTML: An approach to consolidation. In: Proceedings of the RPI Annual Computer Science Conference; CT, USA: Rensselaer Polytechnic Institute Press; 17 December 1996

[5] Mandeville J et al. GreenSpace: Creating a distributed virtual environment for global applications. In: Proceedings of IEEE Networked Reality Workshop; Cambridge, MA, USA: MIT Media Lab Press; 26-28 October 1995

[6] Schwartz P et al. Virtual playground: Architectures for a shared virtual world. In: Proceedings of ACM Virtual Reality Software and Technology Conference (VRST). New York, NY, USA: Association for Computing Machinery; 2-5 November 1998

[7] Campbell B, Rosse C, Brinkley J. The virtual anatomy lab: A hands-on anatomy learning environment. In: Proceedings of the Medicine Meets Virtual Reality Conference (MMVR). San Francisco, CA, USA: IOS Press; 22-24 January 2001

[8] Campbell B et al. Web3D in ocean science learning environments: Virtual big beef creek. In: Proceedings of the Web3D Symposium. Tempe, Arizona, USA: ACM Press; 24-28 February 2002

[9] Damer B et al. OWorld: Connecting 3-D cyberspace. In: Proceedings of the Digital Biota III Conference. Boulder Creek, CA, USA: Digitalspace Press; 6-7 November 1999

[10] Leonardo L. Using Netscape Liveconnect. Hoboken: Que Publishing; 1997. ISBN: 9780789711717

[11] United States. Executive Office of the President and United States. Assistant to the President for Homeland Security and Counterterrorism. The Federal Response to Hurricane Katrina: Lessons Learned. White House: Washington, D.C; 2006

[12] Campbell B, Weaver C. RimSim response hospital evacuation: Improving situation awareness and insight through serious games play and analysis. International Journal of Information Systems for Crisis Response and Management. 2011;**3**(3):1-15

[13] Campbell B, Schroder K, Weaver C. RimSim visualization: An interactive tool for post-event sense making of a first response effort. In: Proceedings of the 7th International Conference on Information Systems for Crisis Response and Management (ISCRAM). New York, NY, USA: Plenum Publishing; 2-5 May 2010

[14] Campbell B, Schroder K. Training for emergency response with RimSim:Response! In: Proceedings of SPIE Defense, Security + Sensing Conference. Bellingham, WA, USA: SPIE Press; 13-17 April 2009

[15] Campbell B et al. Emergency response planning and training

through interactive simulation and visualization with decision support. In: IEEE International Conference on Technologies for Homeland Security. New York, NY, USA: IEEE Publishing; 22-24 May 2008

[16] Campbell B. Adapting simulation environments for emergency response planning and training [doctoral thesis]. Seattle: University of Washington; 2010

[17] three.js [Internet]. 2020. Available from: http://threejs.org [Accessed: December 28, 2021]

[18] Marcos D. A-FRAME: A web framework for building 3D/AR/VR experiences [Internet]. Available from: http://aframe.io [Accessed: December 28, 2021]

[19] Fretin V. Networked-aframe GitHub repository [Internet]. Available from: https://github.com/networked-aframe/networked-aframe [Accessed: December 28, 2021]

[20] The Khronos Group. OpenXR: Unifying reality [Internet]. Available from: https://www.khronos.org/api/index_2017/openxr [Accessed: December 28, 2021]

[21] Anderson S, Jamniczky H, Krigolson O. Quantifying two-dimensional and three-dimensional stereoscopic learning in anatomy using electroencephalography. npj Science of Learning. 2019;**4**:10. DOI: 10.1038/s41539-019-0050-4

[22] Giglioli I et al. Are 3D virtual environments better than 2D interfaces in serious games performance? An explorative study for the assessment of executive functions. Applied Neuropsychology. Adult. 2021;**28**(2): 148-157. DOI: 10.1080/23279095.2019.1607735